H. Bauer H.-D. Klenk Ch. Scholtissek (Eds.)

# Modern Trends in Virology

With 63 Figures and 21 Tables

Springer-Verlag Berlin Heidelberg New York
London Paris Tokyo

Professor Dr. med. HEINZ BAUER
Justus-Liebig-Universität Giessen
Institut für Medizinische Virologie
Frankfurter Strasse 107, D-6300 Giessen

Professor Dr. med. HANS-DIETER KLENK
Philipps-Universität Marburg
Zentrum für Hygiene und Medizinische Mikrobiologie
Institut für Virologie
Robert-Koch-Strasse 17, D-3550 Marburg/Lahn

Professor Dr. rer. nat. CHRISTOPH SCHOLTISSEK
Justus-Liebig-Universität Giessen
Institut für Virologie
Fachbereich Veterinärmedizin
Frankfurter Strasse 107, D-6300 Giessen

ISBN-13:978-3-642-73747-3 e-ISBN-13:978-3-642-73745-9
DOI: 10.1007/978-3-642-73745-9

Library of Congress Cataloging in Publication Data
Modern trends in virology / H. Bauer, H.-D. Klenk, Ch. Scholtissek (eds.). p. cm. Based on a symposium held in Giessen in June 1986, dedicated to Rudolf Rott. Includes bibliographies and index.ISBN-13:978-3-642-73747-3(U.S.)1.Virology–Congresses.2.Medical virology–Congresses. I. Bauer, H. (Heinz), 1933– . II. Klenk, H.D. (Hans-Dieter), 1938– . III. Scholtissek, Ch. (Christoph), 1929– . IV. Rott, R. (Rudolf), 1926– . [DNLM: 1. Viruses–congresses. QW 160 M689 1986] QR355.M63 1988 576'.64–dc19 DNLM/DLC for Library of Congress 88-24935 CIP


Softcover reprint of the hardcover 1st edition 1988



2123/3145-543210 – Printed on acid-free paper

# Table of Contents

# List of Contributors*



* The address of the authors is given on the first page of each contribution
[1] Page, on which contribution commences

# Preface

In June 1986 a symposium was held in Giessen on Modern Trends in Virology. It was initiated by the Deutsche Forschungsgemeinschaft, which had supported virus research for the past 18 years in the Sonderforschungsbereich 47 at the University of Giessen. The purpose of the meeting was to serve as a forum for the members of the Sonderforschungsbereich to discuss scientific topics of mutual interest with about 200 virologists that had come from various parts of Europe, the United States, and Japan. It was not by chance that the symposium took place shortly after the 60th birthday of Rudolf Rott, who had founded the Sonderforschungsbereich in 1968 and has been its speaker ever since. Without his vision and his never resting energy Giessen would not have gained the position in the field of virology that it has today. This Festschrift, which contains the contributions presented at the plenary sessions of the symposium, is therefore dedicated to Rudolf Rott.

HEINZ BAUER
HANS-DIETER KLENK
CHRISTOPH SCHOLTISSEK

# A Genetic Approach to Determining Glycoprotein Topology: The Influenza B Virus NB Glycoprotein has an Extracellular $NH_2$-Terminal Domain Containing two N-linked Carbohydrate Chains*

ROBERT A. LAMB and MARK A. WILLIAMS

Department of Biochemistry, Molecular Biology and Cell Biology, Northwestern University, Evanston, IL 60208, USA

## Introduction

In eukaryotic cells, glycoproteins are either translocated completely across the membrane into the lumen of the endoplasmic reticulum for soluble proteins destined for secretion, or they are asymmetrically integrated into distinct cellular membranes. Integral membrane proteins may span the membrane once or more with the N- and C-termini of the polypeptide chain on either side of the bilayer. These proteins are anchored in membranes by a membrane-spanning domain which is generally found to consist of 20 or more hydrophobic or uncharged amino acids. It has been established that the orientation of an integral membrane protein is absolute and is maintained regardless of the final destination of the protein (e.g., in the plasma membrane) (reviewed in Blobel 1980: Wickner and Lodish 1985). Similarly, the addition of asparagine (N)-linked carbohydrate to an integral membrane protein is also asymmetric and occurs in the lumen of endoplasmic reticulum vesicles at the amino acid sequence Asn-X-Ser/Thr (reviewed in Hubbard and Ivatt 1981). However, determination of the transmembrane orientation and location of the precise sites used for addition of carbohydrate by peptide analysis are often difficult.

We have been interested in examining properties of a glycoprotein, NB, of influenza B virus which circumstantial evidence indicated was an integral membrane protein. In the course of these studies we have developed a method, based on site-specific mutagenesis and expression of mutant cDNAs in eukaryotic cells, that enables the orientation of the protein in membranes and the precise sites of carbohydrate addition to be determined.

Influenza B virus NB glycoprotein (100 amino acids) is translated from a bicistronic mRNA derived from RNA segment 6 that also encodes the neuraminidase (NA) glycoprotein using overlapping reading frames (Shaw et al. 1983; Lamb et al. 1983). The first AUG codon in the mRNA is used for the initiation of NB and the second AUG codon, separated from the first AUG codon by four nucleotides, is used to initiate NA. NB ($M_r$ ~18 000) has been shown to be abundantly expressed in influenza B virus-infected cells and shown to be glycosylated based on incorpora-

* This research was supported by National Institutes of Health research awards AI-20 201 and AI-23 173 and a Grant-in-Aid from the American Heart Association. R.A.L. is an established investigator of the American Heart Association.

tion of [$^3$H]-glucosamine and a shift in apparent $M_r$ after treatment with endoglycosidase H from ~18 000 to ~11 500 (Shaw et al. 1983). Examination of the predicted amino acid sequence of NB (Fig. 1; data derived from Shaw et al. 1982, 1983) indicates that it contains a region of 22 uncharged amino acids (residues 19 to 40). This region has a hydropathic index (Kyte and Doolittle 1982) of >2, a value normally found for regions of proteins that interact with membranes (Kyte and Doolittle 1982; see Fig. 2), but biochemical evidence that NB is an integral membrane-

20
MET-Asn-Asn-Ala-Thr-Phe-Asn-Cys-Thr-Asn-Ile-Asn-Pro-Ile-Thr-His-Ile-Arg- Gly-Ser-Ile-Ile
1 2

40
Ile-Thr-Ile-Cys-Val-Ser-Leu-Ile-Val-Ile-Leu-Ile-Val-Phe-Gly-Cys-Ile-Ala - Lys-Ile-Phe-Ile-

60
Asn-Lys-Asn-Asn-Cys-Thr-Asn-Asn-Val-Ile-Arg-Val-His-Lys-Arg-Ile-Lys-Cys-Pro-Asp-Cys-Glu-
3

80
Pro-Phe-Cys-Asn-Lys-Arg-Asp-Asp-Ile-Ser-Thr-Pro-Arg-Ala-Gly-Val-Asp-Ile-Pro-Ser-Phe-Ile-

100
Leu-Pro-Gly-Leu-Asn-Leu-Ser-Glu-Gly-Thr-Pro-Asn
4

**Fig. 1.** The amino acid sequence of B/Lee/40 NB predicted from the nucleotide sequence of the NA/NB mRNA (data derived from Shaw et al. 1983). The four potential glycosylation sites are *underlined and numbered* with respect to their position from the N-terminus, the long uncharged hydrophobic domain is *boxed* and the oligopeptide used to make an antibody to NB is indicated by *double underlines.* (From Williams and Lamb 1986)

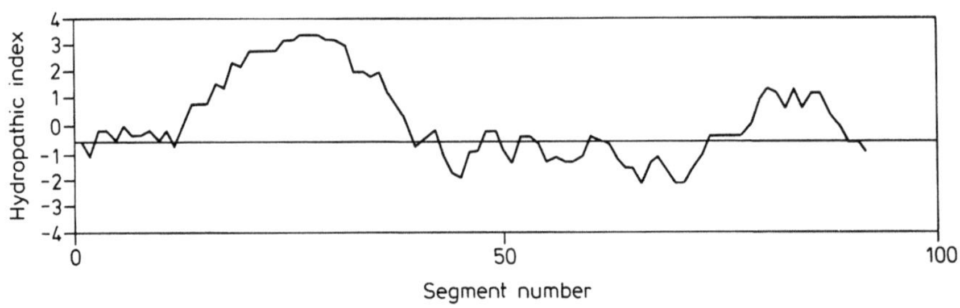


**Fig. 2.** Hydropathy plot of the NB protein. The relative hydrophobicity and hydrophilicity of the protein along its sequence was calculated as described (Kyte and Doolittle 1982) using a segment length of nine amino acids. The consecutive scores are plotted from the N- to the C-terminus of the protein, and the midpoint line corresponds to the grand average of hydropathy of the amino acid compositions found in most sequenced proteins. (Kyte and Doolittle 1982)

protein has not been reported and the subcellular localization of NB has not been determined. NB contains four potential sites (Asn-X-Ser/Thr) for the addition of N-linked carbohydrate, two on each side of the hydrophobic domain. As discussed above, the actual sites used depend on the orientation of the protein in membranes. We present both biochemical and molecular-genetic evidence that NB is an integral membrane protein expressed at the cell surface with an N-terminal region and two carbohydrate chains exposed extracellularly, an orientation unusual for a protein lacking a cleavable signal sequence.

## Experimental Procedures

For details of the methods used the reader is referred to Williams and Lamb (1986).

## Results

### Production of Antisera Specific to NB

To facilitate experiments and to confirm indirectly a portion of the sequence of NB, two antisera were produced that are specific for NB. The predicted amino acid sequence of NB (Fig. 1) was used to synthesize an oligopeptide from a C-terminal region (CNB, residues 58–72). Rabbit "site-specific" antisera were prepared to the oligopeptide coupled to keyhole limpet hemocyanin. The second antiserum (DNB) was produced in rabbits to gel-purified and denatured NB. The specificity of the two antisera for NB was tested by immunoprecipitation of both [$^{35}$S]-cysteine-labeled influenza B/Lee/40 virus-infected CV1 and MDCK cell extracts and is shown in Fig. 3. It can be seen that the DNB and CNB antisera readily precipitate NB (lanes D and F, J and L respectively) and that these antisera do not precipitate any polypeptides from uninfected cells similar in mobility to NB. In samples immunoprecipitated with the DNB and CNB antisera, bands similar in mobility to $NS_1$ and NP are also observed. This precipitation is probably artifactual and does not affect interpretation of the data presented here. In the immunoprecipitates from infected CV1 cell lysates a region of diffuse radioactive material is observed migrating slower than NB, and as discussed later (see Fig. 7), this is probably caused by carbohydrate side chain heterogeneity on NB such as extensions on the N-linked glycans. In the immunoprecipitates from infected MDCK cells a region of diffuse radioactive material is observed migrating between NP and $NS_1$ (Fig. 3, lanes J and L), and this is discussed below (see Fig. 4).

### NB Is Expressed at the Cell Surface with an Extracellular N-terminus and Cytoplasmic C-terminus

The data described above suggest that NB is an integral membrane glycoprotein anchored in membranes by the single hydrophobic region located near the N-terminus. To examine whether NB is expressed at the infected cell surface and to deter-

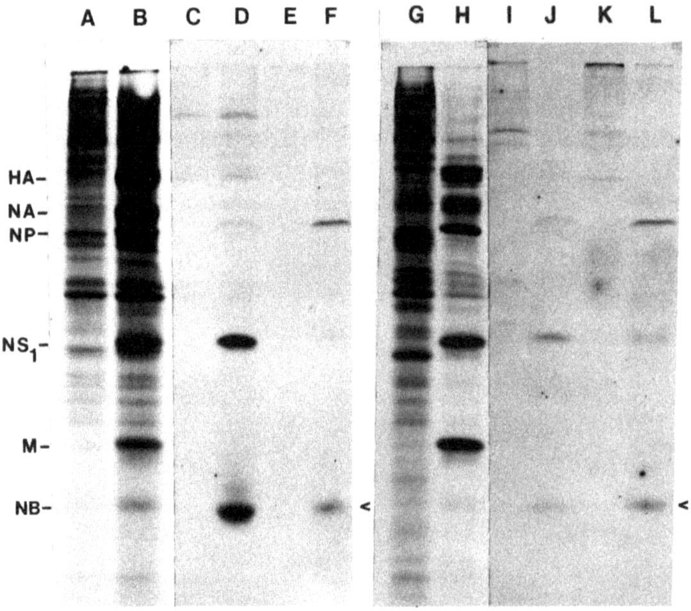


**Fig. 3.** Production of antisera to NB. Influenza B/Lee/40-infected CV1 (lanes *A–F*) and MDCK (lanes *G–L*) cells were labeled with 50 μCi [$^{35}$S]-cysteine at 8 h p.i. for 2 h. Immunoprecipitations were done as described by Williams and Lamb (1986). Lanes *A* and *G*, mock-infected cells; lanes *B* and *H*, infected cells; lanes *C* and *I*, immunoprecipitation of mock-infected cell lysates with DNB antiserum; lanes *D* and *J*, immunoprecipitation of infected cell lysates with DNB antiserum; lanes *E* and *K*, immunoprecipitation of mock-infected cell lysates with CNB antiserum; lanes *F* and *L*, immunoprecipitation of infected cell lysates with CNB antiserum. Molecular weight markers for all polyacrylamide gels are influenza B/Lee/40 infected cell polypeptides of the following $M_r$: HA, 82000; NA, 66000; NP, 63000; NS, 40000; M, 24000; NB, 18000. (From Williams and Lamb 1986)

mine the topography of NB in membranes, intact cells were treated with proteinase K. [$^{35}$S]-Cysteine-labeled infected MDCK cells were gently scraped from tissue culture plates and incubated with 100 μg/ml proteinase K for 40 min at 37 °C. It was found that most of hemagglutinin (HA) and NA were selectively removed, as would be expected for surface spike glycoproteins (Fig. 4, compare lanes 1 and 2), and that viral proteins known to reside inside the cell (e.g., NP, $NS_1$ and M) were largely protected from protease digestion. A decrease in the amount of NB was observed with the appearance of several new bands (Fig. 4, compare lanes 1 and 2). To determine whether any of the new bands were related to NB, the samples were immunoprecipitated with the "site-specific" C-terminal CNB antisera. After proteinase K treatment of the cell surface some NB was found in a "trimmed" form (Fig. 4, lane 6, band d) of $M_r$ ~10000. The large shift in mobility from NB (band a) to "trimmed" NB (band d) would be expected if proteinase K removed a peptide containing carbohydrate residues. These data would suggest that some of the NB molecules are accessible to digestion with proteinase K at the cell surface, and because of the specificity of the CNB antisera, make it likely that the N-terminus of NB, containing the sugar chains, is exposed at the cell surface. To attempt to obtain a com-

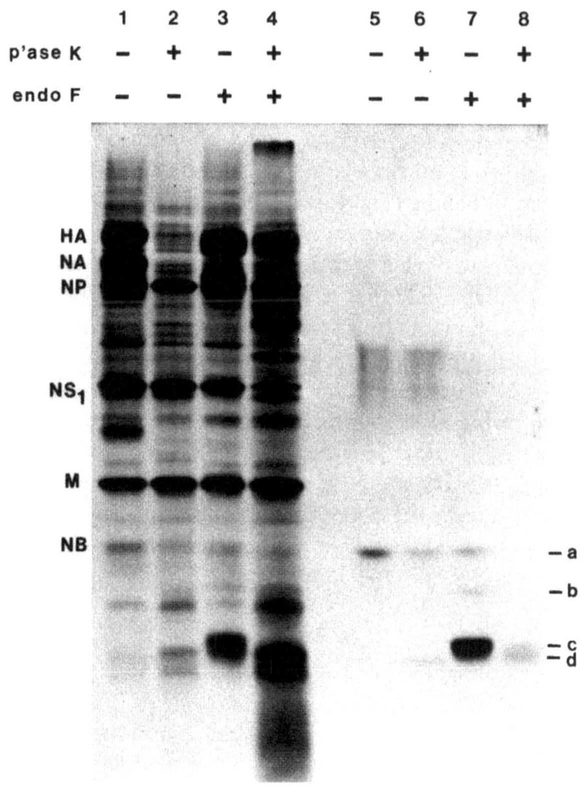


**Fig. 4.** Evidence for the expression of NB on the cell surface with its N-terminus and carbohydrate chains exposed. MDCK cells were infected with influenza B/Lee/40, labeled with [$^{35}$S]-cysteine, scraped from plates and resuspended in PBS. The intact cells were untreated (−) or treated (+) at 37 °C with the enzymes indicated above the lanes: *p'ase K,* 100 μg/ml proteinase K for 40 min; *endo F,* 1 U endoglycosidase F for 3 h. Lanes *1–4* are direct lysates of the cells. Lanes *5–8* show the immunoprecipitation with CNB peptide antibody of the samples shown in lanes 1–4. Protein bands immunoprecipitated by CNB antibody are denoted by lowercase letters (*a–d*) and are discussed in the text. The amounts of lysates used were adjusted to make the level of NS protein approximately equivalent, to correct for cells lost in the washing procedure, except for the immunoprecipitation in lane 8 where the available cells were used. (From Williams and Lamb 1986)

plete conversion of NB to the "trimmed" form, higher concentrations of protease and longer incubation times were used but these resulted in unacceptable levels of cell lysis.

As an alternative approach to show that NB is expressed at the cell surface, intact cells were treated with endoglycosidase F (endo F) to examine for the loss of carbohydrate from NB. As shown in the direct lysate (Fig. 4, lane 3) and in lysates immunoprecipitated with CNB antisera (Fig. 4, lane 7), two new bands were observed; a minor component band b ($M_r$ ~15 000) and a major component band c ($M_r$ ~11 500). Band c has identical mobility to that observed for NB immunoprecipitated from infected cell lysates and treated with endo F (data not shown). Band b is thought to result from incomplete digestion with the enzyme and probably represents a form of NB that has lost one carbohydrate chain (see Fig. 7). The reproducible large increase in amount of radioactivity in NB band c after endo F treatment of the cell surface was unexpected, but in influenza B virus-infected MDCK cells we have consistently observed a diffuse region of bands ($M_r$ ~35000–50000) (Fig. 4, lanes 5 and 6) after precipitation with CNB or DNB antisera. These bands are lost on endo F treatment of the cell surface, and concomitantly a large increase in the amount of band c (Fig. 4, lane 7) is observed. The CNB antisera does not precipitate any polypeptides from uninfected cells treated with endo F (not shown), eliminating the unlikely possibility that endo F treatment of cells reveals an antigenic epitope of a host polypeptide that is recognized by the "site-specific" CNB antisera and also has the same mobility as endo F-treated NB. Thus, it seems likely that NB interacts with itself or other molecules through the MDCK cell-specific carbohydrate moieties.

To further examine the N-terminal region of NB exposed at the cell surface we attempted to treat intact cells sequentially with endo F and proteinase K. These treatments make the cells very fragile and it is difficult to avoid considerable losses of cells during the procedure. Nonetheless, by selecting endo F and proteinase K conditions we were able to detect some remaining NB-specific bands c (Fig. 4, lanes 4 and 8).

## Expression of NB and NA from an SV40 Recombinant Virus

To further examine the synthesis and transport of NB to the plasma membrane and to provide a means of examining the phenotype of genetically engineered mutants in NB, the cDNA to influenza B virus RNA segment 6 was expressed in eukaryotic cells using an SV40 late-region replacement vector. In influenza B virus-infected cells, both NB and NA are translated from a bicistronic mRNA derived from RNA segment 6 (Shaw et al. 1983). Therefore, it was expected that both NB and NA would be expressed when the cDNA copy of RNA segment 6 was transcribed under the control of the SV40 late-region promoter and polyadenylation signal. The vector used, SV93, is very similar to SV103, described previously (Paterson et al. 1985), except for restriction endonuclease sites available for cloning insert cDNAs. The G/C tails were removed from the 5′ end (mRNA sense) of the cDNA of RNA segment 6 (pBNA1, Shaw et al. 1982) by controlled nuclease Bal 31 and restriction endonuclease digestions. Two deletions were used in the experiments described here. pSV93BNA$\Delta$4 contains a deletion of three influenza virus-specific nucleotides and pSV93BNA$\Delta$39 contains a deletion of 38 influenza virus-specific nucleotides, both from the 5′ end of the DNA in the mRNA sense.

The recombinant SV40-B/NB/NA DNA molecules with the B/NB/NA DNA inserted in either the mRNA or viral RNA (vRNA) sense with respect to SV40 late-region transcription were transfected into CV1 cells, together with an SV40 early-region deletion mutant (SVdl1055), and lytic stocks of virus were prepared.

When the SV40-B/NB/NA (mRNA and vRNA sense)-infected CV1 cells were labeled with [$^{35}$S]-methionine at 48 h p.i. for 3 h and immunoprecipitated with NA antiserum (Fig. 5, left five lanes), a band identical in mobility to NA, precipitated from B/Lee/40 virus-infected cells, was detected from the mRNA sense vectors ($\Delta$4

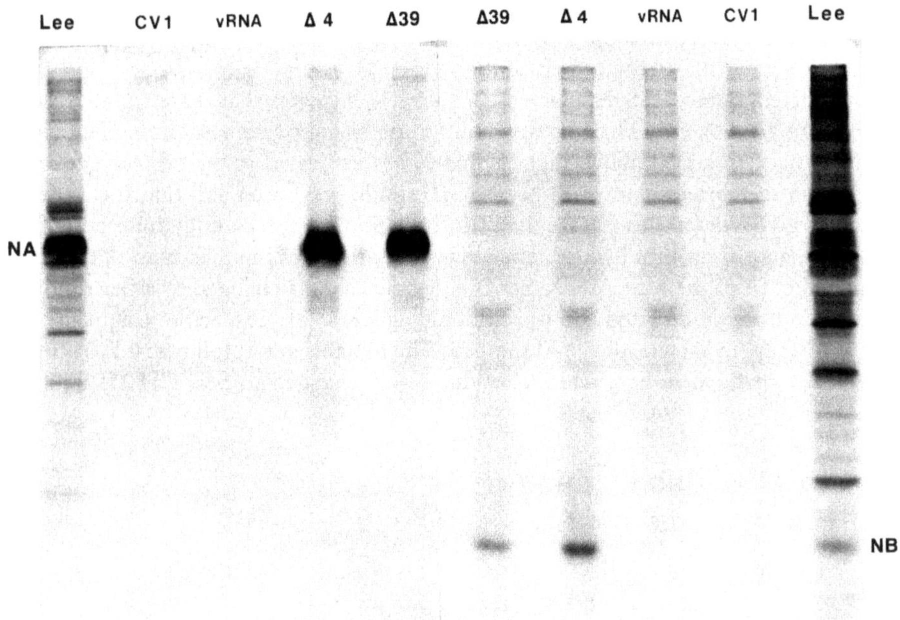


**Fig. 5.** Expression of NA and NB proteins from an SV40 vector. CV1 cells were infected with the SV40 recombinant viruses, labeled with [$^{35}$S]-methionine (*left* five lanes) or [$^{35}$S]-cysteine (*right* five lanes) and immunoprecipitated. The five lanes on the left side of the figure were immunoprecipitated with NA antibody. The lanes on the right side of the figure (with the exception of the last lane) were immunoprecipitated with DNB antibody. Lane Lee is a direct lysate. Lee, cells infected with B/Lee/40; CV1, mock infected CV1 cells; vRNA, SV93BNA$\Delta$4 vector with B/NB/NA DNA insert in the missense orientation with respect to SV40 late-region transcription; $\Delta$4, SV93BNA$\Delta$4 recombinant virus; $\Delta$39, SV93BNA$\Delta$39 recombinant virus. $\Delta$4 and $\Delta$39 contain the B/NB/NA DNA in the mRNA sense with respect to SV40 late-region transcription. (From Williams and Lamb 1986)

and $\Delta$39) but not from the vRNA sense vector or from uninfected cells. To examine for the synthesis of NB, the vector-infected CV1 cells were labeled at 48 h p.i. for 3 h with [$^{35}$S]-cysteine and immunoprecipitated with DNB antiserum. As shown in Fig. 5 (right five lanes), a band identical in mobility to NB from influenza virus-infected cells was detected from the mRNA sense vectors ($\Delta$4 and $\Delta$39) but not from the vRNA sense vector or from uninfected cells, indicating that NB was expressed from the vector and glycosylated normally. It has been reported that NB has a short half-life (less than 1 h) in influenza B/Lee/40-infected HeLa cells (Shaw and Choppin 1984), but in influenza B virus-infected CV1 and MDCK cells NB has a half-life of 4–5 h (data not shown), and NB does not turn over appreciably during the 3-h labeling period used in these SV40 recombinant virus experiments. In influenza B virus-infected cells, NB and NA are synthesized in approximately equimolar amounts (Shaw et al. 1983). The relative amounts of NA and NB shown in Fig. 5 do not directly reflect amounts of NA and NB synthesized by the vector, as different radioactive isotopes and antisera of different titers were used.

## Identification of the Precise Carbohydrate Sites in NB Using Site-Specific Mutagenesis and Expression in Eukaryotic Cells

The predicted amino acid sequence of NB (Fig. 1) indicates that there are four potential sites (Asn-X-Ser/Thr) for the addition of N-linked carbohydrate. The biochemical data described above suggest that NB is expressed at the cell surface and oriented in membranes with ~18 N-terminal residues exposed and that this region of NB is glycosylated. To obtain direct information on the carbohydrate addition sites used, and to confirm by a genetic means the orientation in membranes, the Asn residue in each of the four Asn-X-Ser/Thr sequences was changed to a Ser residue by oligonucleotide-directed site-specific mutagenesis of a specific fragment of pSV93BNA$\Delta$4 in bacteriophage M13mp19. The mutants are as follows: CH01, residue 3 Asn changed to Ser; CH02, residue 7 Asn changed to Ser; CH03, residue

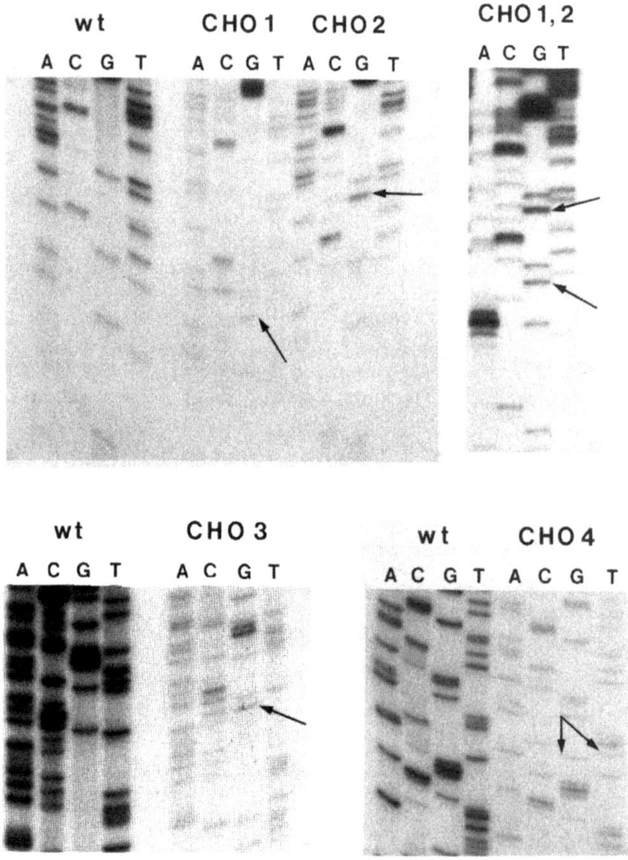


**Fig. 6.** NB glycosylation mutant sequences. The nucleotide sequences of each NB glycosylation mutant plasmid DNA and the wild type *(wt)* in the corresponding regions are shown, except CHO 1,2, which shows the sequence of the single-stranded M13 bacteriophage DNA. The *lanes* are labeled to indicate the specific bases, and bases changed by site-specific mutagenesis are indicated by *arrows*

48 Asn changed to Ser; CH04, residue 93 Asn changed to Ser; CH01,2, a double mutant in which the Asn at residues 3 and 7 were both changed to Ser. The nucleotide sequences of the mutants are shown in Fig. 6. The mutants, once constructed into the SV40 vector described above, were expressed in CV1 cells. The identification of the mutants was expected to be unambiguous due to the large difference in mobility between NB ($M_r$ ~18000) and endo F-treated NB ($M_r$ ~11500) (see Fig. 4). The phenotype of the mutants that had "lost" a glycosylation site was predicted to be an NB-specific polypeptide of intermediate gel electrophoretic mobility (i.e., $M_r$ ~14000–15000).

SV40 recombinant virus-infected CV1 cells were labeled with [$^{35}$S]-cysteine and immunoprecipitated with CNB antibody. NB ($M_r$ ~18000) expressed from the wild-type SV40BNAΔ4 vector is indicated by an arrow in Fig. 7, lane A. NB expressed from the wild-type vector was treated with endo F and immunoprecipitated from cell lysates. The mobility of NB was shifted to $M_r$ ~11500 (Fig. 7, lane B) as observed after endo F treatment of influenza virus-infected cells (Fig. 4). The mutants CH01 and CH02 expressed major new NB-specific bands of $M_r$ ~14000–15000 (lanes 1 and 2 respectively) whereas mutants CH03 and CH04 expressed major NB-specific bands of $M_r$ ~18000 (lanes 3 and 4 respectively). The

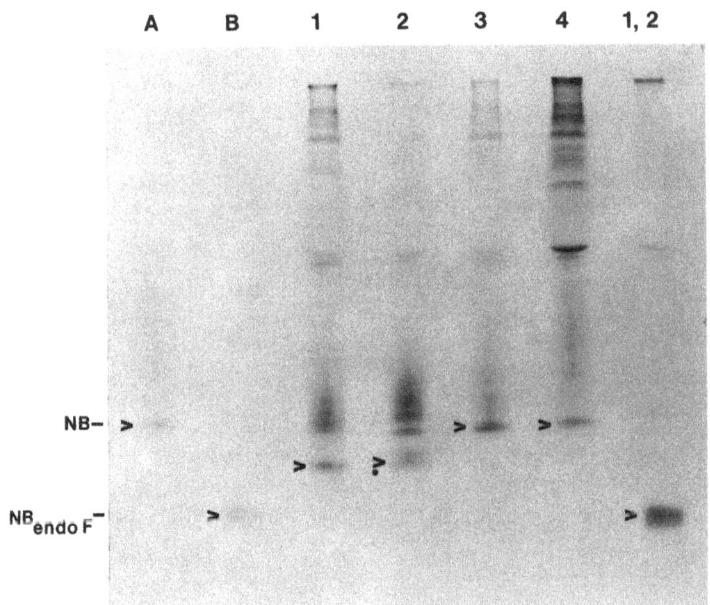


**Fig. 7.** Expression of NB glycosylation mutants in an SV40 recombinant vector. CV1 cells were infected with the SV40 recombinant viruses containing the glycosylation site mutants in NB, labeled with [$^{35}$S]-cysteine at 48 h p.i., lysed and immunoprecipitated with CNB antibody. Lane *A*, wild-type NB expressed from SV93BNAΔ4; lane *B*, wild-type NB expressed from SV93BNAΔ4 and the cell lysate treated with 0.25 units endoglycosidase F *(endo F)* prior to immunoprecipitation; lanes *1–4*, expression of the mutant NB from the vectors containing the CHO1, CHO2, CHO3 and CHO4 mutations respectively; lane *1,2*, expression of NB from the double mutation CHO1,2. (From Williams and Lamb 1986)

double mutant CH01,2 expressed a major NB-specific band of $M_r$ ~10500. Thus, these data would indicate that the N-terminal Asn residues 3 and 7, which are exposed at the cell surface, are both used for the addition of carbohydrate. It can be seen in Fig. 7, lanes 1 and 2 that the mutants CH01 and CH02 express additional NB-related species not seen with the wild-type vector, or seen in much less abundance. One of the additional bands migrates with a mobility similar but not identical to that of glycosylated wild-type NB. To rule out the trivial explanation that these species had arisen because of contamination of the mutants, the nucleotide sequence of NB in the vector DNA molecules was obtained immediately prior to transfection into the cells. Additionally, the mutation in CH01 eliminates the AUG codon for NA (see Fig. 6). No expression of NA was detected with CH01 at a level that would have detected 1% wild-type expression. The observation that these additional bands are lacking from mutant CH01,2 (no carbohydrate chains) suggests that they arise by heterogeneity in the composition of the carbohydrate chains, especially when there is only one sugar chain attached to NB, which leads to different electrophoretic forms on gel analysis. We found further that the extra NB-specific bands and changes in the electrophoretic mobility of NB in CH01, CH02 and CH01,2 were consequences of the lack of one or two carbohydrate chains. The NB-specific proteins expressed by the vectors were treated with endo F and precipitated with CNB

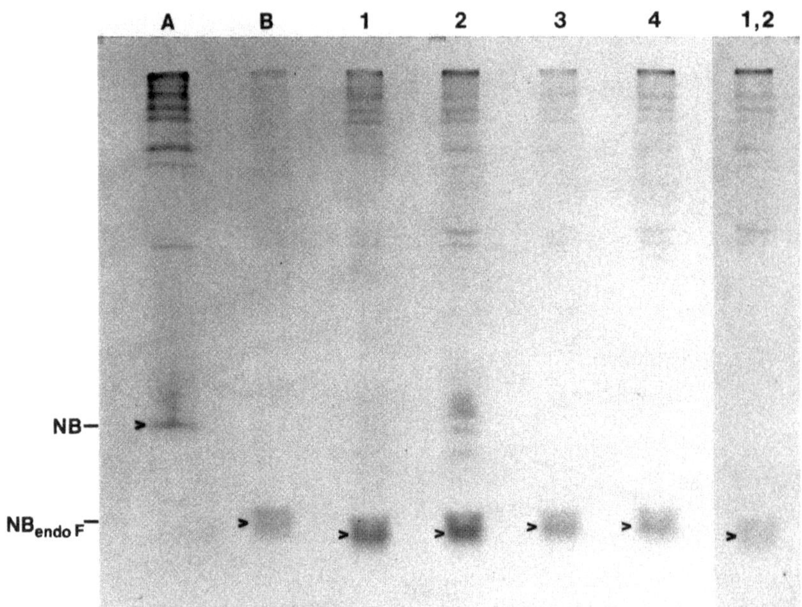


**Fig. 8.** Endoglycosidase F *(endo F)* treatment of NB expressed by the CHO mutants. CV1 cells were infected with the SV40 recombinant vectors, labeled with [$^{35}$S]-cysteine and lysed as described above. Aliquots of each lysate were treated with 0.25 units endo F for 12 h at 37 °C prior to immunoprecipitation. Lane *A*, wild-type untreated NB expressed from SV93BNA$\Delta$4; lane *B*, wild-type NB expressed from SV93BNA$\Delta$4; lanes *1–4*, expression of the mutant NB from the vectors containing CHO1, CHO2, CHO3 and CHO4 mutations respectively; lane *1,2*, expression of NB from the double mutation CHO1,2. (From Williams and Lamb 1986)

antibody. As shown in Fig. 8, a single major species of NB ($M_r$ ~11 500) was observed in all the mutants, except for CH02 where the digestion with endo F is thought to be incomplete. One possible explanation for the heterogeneity of NB mobility observed for CH01 and CH02 (Fig. 7) is sulfation of the carbohydrate, a modification known to occur for HA and NA of influenza virus (Nakamura and Compans 1979). The degree of sulfation may lead to different electrophoretic forms on SDS polyacrylamide gels. A precedent for this type of phenomenon is found in many phosphoproteins, where multiple electrophoretic bands are found on gels depending on the level of phosphorylation, e.g., M of Sendai virus (Lamb and Choppin 1977) and NS of vesicular stomatitis virus (VSV) (Kingsford and Emerson 1980). It can be seen in Fig. 8 that there is a small variation in the mobility of NB after endo F treatment between the mutants and the wild-type NB. This is not unexpected, as in each mutant an Asn residue has been changed to Ser and it is well established, especially with influenza virus polypeptides (reviewed in Lamb 1983), that single amino acid changes can slightly alter the mobility of a polypeptide on gels.

## Discussion

The data presented here indicate that NB (100 amino acids), the third glycoprotein of influenza B virus in addition to HA and NA, is an integral membrane protein expressed at the infected cell surface. A single hydrophobic domain (residues 19–40) anchors the protein in membranes, and ~18 N-terminal amino acids containing two endo F-sensitive carbohydrate chains are exposed at the cell surface. A schematic diagram of the structure of NB is shown in Fig. 9. This structure was demonstrated by biochemical methods and the data were complemented by a genetic approach in-

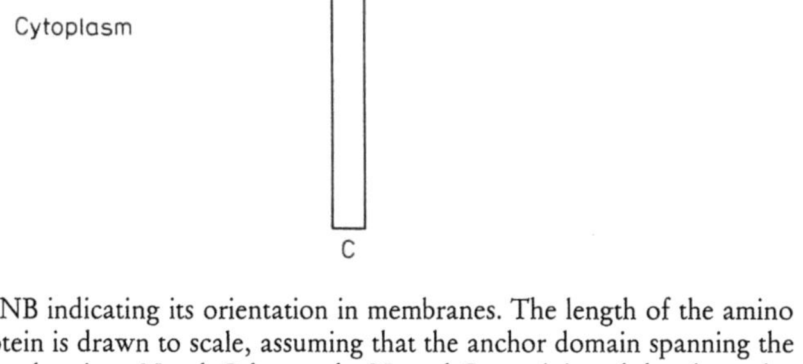


**Fig. 9.** Schematic diagram of NB indicating its orientation in membranes. The length of the amino acid backbone of the glycoprotein is drawn to scale, assuming that the anchor domain spanning the membrane is the entire uncharged region. *N* and *C* denote the N- and C-termini, and the glycosylation sites shown to be used are indicated at their positions relative to the N-terminus. (From Williams and Lamb 1986)

volving deletion of potential glycosylation sites in the NB cDNA using site-specific mutagenesis and expression of the altered proteins in cells. The gel mobility differences among the expressed polypeptides allowed the identification of the precise carbohydrate sites used for glycosylation of NB (Asn at residues 3 and 7). This technique should be generally applicable whenever a cDNA clone of a transmembrane protein is available and should be particularly useful in determining the glycosylated domains exposed extracellularly of proteins that span the membrane more than once. The well-characterized N-linked carbohydrate addition sites of the VSV G protein have also been deleted using site-specific mutagenesis (Machamer et al. 1985), but the purpose of these experiments was to determine the role of sugar chains on transport of G to the cell surface.

The role of NB in influenza B virus infections is not known. Although NB is readily detected in infected cells, it has not been detected in purified virions, but this may be dependent on the sensitivity of the methods employed (Shaw et al. 1983). The absence of NB in any great amount from virions as compared to HA and NA suggests that there must be an exclusion mechanism for NB before or during the budding process. It will be of great interest to determine the geographic distribution of NB on the cell surface with respect to budding virions. The cell surface localization of NB makes it unlikely that the protein is involved in transcription or replication of RNA. A role in organizing proteins at the cell surface to form patches of viral proteins or in the budding process would seem more likely.

Influenza A and B viruses both contain eight RNA segments, and at least nine polypeptides encoded by influenza B viruses have counterparts in influenza A virus: PB1, PB2, PA, HA, NA, NP, $M_1$, $NS_1$ and $NS_2$ polypeptides (reviewed in Lamb 1983). A major difference between influenza A and B viruses is the finding of the NB glycoprotein in influenza B viruses. NB is translated from a bicistronic mRNA derived from RNA segment 6 that also encodes NA using overlapping reading frames (Shaw et al. 1983). A second open reading frame has not been found on RNA segment 6 of influenza A viruses. However, influenza A viruses encode the $M_2$ polypeptide that is translated from a spliced mRNA derived from RNA segment 7 and overlaps the region encoding $M_1$ (Lamb et al. 1981). No counterpart for $M_2$ has been reported to be found in influenza B virus-infected cells, although RNA segment 7 of influenza B virus, which encodes the $M_1$ protein, does contain a second overlapping reading frame (Briedis et al. 1982). Recently, it has been shown that influenza A virus $M_2$ protein is an integral membrane protein expressed at the infected cell surface (Lamb et al. 1985). The results presented here for NB of influenza B virus show that it has structural similarities with $M_2$ of influenza A virus. These include the following: (1) NB and $M_2$ are of similar sizes (100 and 97 amino acids respectively) and contain a single internal hydrophobic domain. (2) NB and $M_2$ are integral membrane proteins and localized on the infected cell surface. (3) NB and $M_2$ are oriented with a short N-terminal region exposed at the cell surface and a longer C-terminal cytoplasmic domain (~60 amino acids for NB and ~53 amino acids for $M_2$). (4) NB and $M_2$ are greatly underrepresented or absent from purified virions. However, NB and $M_2$ do not share primary sequence homology, and NB is glycosylated whereas $M_2$ is not (Zebedee et al. 1985). Whether the similarity of structural features between NB and $M_2$ is fortuitous or reflects a common function is not known.

Integral membrane proteins are found in several topographical forms with respect to the lipid bilayer. Many viral and cellular glycoproteins have a cleaved N-terminal signal sequence, span the membrane once and are oriented with a large N-terminal extracellular domain and a small C-terminal cytoplasmic tail, e.g., influenza virus HA, paramyxoviruses F, VSV G and Semliki forest virus $E_1$ and $E_2$ (reviewed in Wickner and Lodish 1985). Some glycoproteins have a single hydrophobic domain (extended signal anchor) in an N-terminal region, span the membrane once and have a large C-terminal extracellular domain and a small N-terminal cytoplasmic tail, e.g., influenza virus NA, paramyxoviruses HN and human asialoglycoprotein receptor (reviewed in Wickner and Lodish 1985). Several cellular membrane proteins contain uncleaved signal sequences near the N-terminus and span the membrane several times with cytoplasmic N- and C-termini e.g., human glucose transport protein, erythrocyte band 3, MP26 of the bovine lens fiber membrane, $Ca^{2+}$-$Mg^{2+}$ ATPase and the catalytic subunit of $Na^{+}$-$K^{+}$ ATPase (reviewed in Wickner and Lodish 1985). However, the known examples of proteins with an uncleaved N-terminal region hydrophobic domain that anchors the protein in membranes with an N-terminal extracellular domain and C-terminal cytoplasmic domain are much smaller in number. Influenza B virus NB, influenza A virus $M_2$ (Lamb et al. 1985) and, most likely, avian erythroblastosis virus gp74v-erb B (Yamamoto et al. 1983) have a single N-terminal region hydrophobic domain that anchors the protein in the lipid bilayer. Bovine opsin does not contain a cleavable N-terminal signal, spans the membrane several times and has an extracellular N-terminus. The opsin internal signal sequences interact with the signal recognition particle (SRP) to target and facilitate cotranslational insertion of the nascent polypeptide chain across the rough endoplasmic reticulum membrane (Perara et al. 1986).

It remains to be determined whether the single uncleaved hydrophobic domain of NB (and $M_2$) can be defined as an "extended signal anchor" that interacts with SRP to cotranslationally insert NB into membranes and then acts as a "stop transfer" sequence to anchor NB in membranes. If this is the case it is interesting to note that ~78% of the NB polypeptide chain would be completed before the interaction with SRP occurs, because when the hydrophobic domain is exposed, a further ~40 amino acids are protected by the ribosome (Walter et al. 1984). It is also possible that SRP may interact with the hydrophobic domain post-translationally, as has been suggested for the human glucose transporter protein (Mueckler and Lodish 1986). Alternatively, NB may insert into membranes post-translationally, independently of SRP, with the hydrophobic domain acting as an "insertion sequence" (Anderson et al. 1983).

*Acknowledgements.* We thank Diane Braun and Margaret Shaughnessy for excellent technical assistance, Christopher Richardson for synthesizing the NB oligopeptide and Robert Webster for antisera to NA.

## References

Anderson DJ, Mostov KE, Blobel G (1983) Mechanisms of integration of de novo-synthesized polypeptides into membranes: signal recognition particle is required for integration into microsomal membranes of calcium ATPase and of lens MP26 but not of cytochrome $b_5$. Proc Natl Acad Sci USA 80: 7249–7253

Blobel G (1980) Intracellular protein topogenesis. Proc Natl Acad Sci USA 77: 1496–1500
Briedis DJ, Lamb RA, Choppin PW (1982) Sequence of RNA segment 7 of the influenza B virus genome: partial amino acid homology between the membrane proteins ($M_1$) of influenza A and B viruses and conservation of a second open reading frame. Virology 116: 581–588
Hubbard SC, Ivatt RJ (1981) Synthesis and processing of asparagine-linked oligosaccharides. Annu Rev Biochem 50: 555–583
Kingsford L, Emerson SU (1980) Transcriptional activities of different species of NS protein purified from vesicular stomatitis virus and cytoplasm of infected cells. J Virol 33: 1097–1105
Kyte J, Doolittle RF (1982) A simple method for displaying the hydropathic character of a protein. J Mol Biol 157: 105–132
Lamb RA (1983) The influenza virus RNA segments and their encoded proteins. In: Palese P and Kingsbury DW (eds) Genetics of influenza viruses. Springer, Vienna, pp 21–69
Lamb RA, Choppin PW (1977) The synthesis of Sendai virus polypeptides in infected cells. III. Phosphorylation of polypeptides. Virology 81: 382–397
Lamb RA, Lai C-J, Choppin PW (1981) Sequences of mRNAs derived from genome RNA segment 7 of influenza virus: colinear and interrupted mRNAs code for overlapping proteins. Proc Natl Acad Sci USA 78: 4170–4174
Lamb RA, Shaw MW, Briedis DJ, Choppin PW (1983) The nucleotide sequence of the neuraminidase gene of influenza B virus reveals two operlapping reading frames. In: Laver WG (ed) The origin of pandemic influenza viruses. Elsevier, New York, pp 77–86
Lamb RA, Zebedee SL, Richardson CD (1985) Influenza virus $M_2$ protein is an integral membrane protein expressed on the infected cell surface. Cell 40: 627–633
Machamer CE, Florkiewicz RZ, Rose JK (1985) A single N-linked oligosaccharide at either of the two normal sites is sufficient for transport of vesicular stomatitis virus G protein to the cell surface. Mol Cell Biol 5: 3074–3083
Mueckler M, Lodish HF (1986) The human glucose transporter can insert posttranslationally into microsomes. Cell 44: 629–637
Nakamura K, Compans RW (1979) Host cell- and virus strain-dependent differences in oligosaccharides of hemagglutinin glycoproteins of influenza A viruses. Virology 95: 8–23
Paterson RG, Hiebert SW, Lamb RA (1985) Expression at the cell surface of biologically active fusion and haemagglutinin/neuraminidase proteins of the paramyxovirus simian virus 5 from cloned cDNA. Proc Natl Acad Sci USA 82: 7520–7524
Perara E, Rothman RE, Lingappa VR (1986) Uncoupling translocation from translation: implications for transport of proteins across membranes. Science 232: 348–352
Shaw MW, Choppin PW (1984) Studies on the synthesis of the influenza B virus NB glycoprotein. Virology 139: 178–184
Shaw MW, Lamb RA, Erikson BW, Briedis DJ, Choppin PW (1982) Complete nucleotide sequence of the neuraminidase gene of influenza B virus. Proc Natl Acad Sci USA 79: 6817–6821
Shaw MW, Lamb RA, Choppin PW (1983) A previously unrecognized influenza B virus glycoprotein from a bicistronic mRNA that also encodes the viral neuraminidase. Proc Natl Acad Sci USA 80: 4879–4883
Walter P, Gilmore R, Blobel G (1984) Protein translocation across the endoplasmic reticulum. Cell 38: 5–8
Wickner WT, Lodish HF (1985) Multiple mechanisms of protein insertion into and across membranes. Science 230: 400–407
Williams MA, Lamb RA (1986) Determination of the orientation of an integral membrane protein and sites of glycosylation by oligonucleotide-directed mutagenesis: the influenza B virus NB glycoprotein lacks a cleavable signal sequence and has an extracellular $NH_2$-terminal region. Mol Cell Biol 6: 4317–4328
Yamamoto T, Nishida T, Miyajima N, Kawai S, Ooi T, Toyoshima K (1983) The *erbB* gene of avian erythroblastosis virus is a member of the *src* gene family. Cell 35: 71–78
Zebedee SL, Richardson CD, Lamb RA (1985) Characterization of the influenza virus $M_2$ integral membrane protein and expression at the infected-cell surface from cloned cDNA. J Virol 56: 502–511

# Paramyxovirus Metabolisms Associated with the Cytoskeletal Framework

Yoshiyuki Nagai, Tetsuya Toyoda, and Michinari Hamaguchi

Research Institute for Disease Mechanism and Control, Nagoya University School of Medicine, Nagoya 466, Japan

## Introduction

Virus metabolisms do not occur randomly within a cell but are localized to specific compartments or structures. Recent studies have suggested that virtually all animal viruses utilize the three-dimensional frameworks, cytoplasmic and nuclear skeletons, for at least part of their metabolism (Penman 1985). Even the glycolytic enzymes, which had long been thought to represent a typical soluble complex within a cell, have been shown to be associated with the cytoskeleton (Masters 1984). The findings of these studies may emphasize more positive involvement of cellular frameworks in biological processes than formerly considered.

Newcastle disease virus (NDV), a prototype paramyxovirus, is a negative-strand virus containing a 50S genome RNA in a structure composed of six proteins. The 50S RNA is associated with the major structural protein subunits, the nucleocapsid protein (NP) and two other minor components, P and L, which are required for transcription (Hamaguchi et al. 1983). The membrane (M) protein is external to the nucleocapsid but inside the viral envelope. The hemagglutinin-neuraminidase (HN) and fusion (F) glycoproteins form the outer envelope with a lipid bilayer derived from the host cell. We report results which suggest that the cytoskeletal framework may participate actively in some steps of virus growth, particularly in the formation of nucleocapsid, virus budding, and viral RNA synthesis.

## Nucleocapsid Formation Associated with the Cytoskeletal Framework

BHK-21 cells infected with NDV (multiplicity of infection: 10 plaque-forming units (PFU)/cell) were pulse-labeled for 10 min at 6 h p.i. with [$^{35}$S]-methionine (10 μCi/ml), or chased for 20 and 60 min at this time. The cells were separated into soluble fraction (SOL) and the remnant by exposing them to an extraction buffer (10 m*M* Pipes, pH 6.8, 0.1 *M* KCl, 2.5 m*M* $MgCl_2$, 1 m*M* $CaCl_2$, 0.3 *M* sucrose, 1 m*M* PMSF) containing 0.5% Triton X-100. The remnant was the cytoskeleton (CSK) together with the nuclei. For further studies of CSK-associated proteins, the nuclei were removed by centrifugation after homogenization of the remnant in 0.5% deoxycholate and 1% Tween 40 in reticulocyte standard buffer (10 m*M* Tris-HCl, pH 7.4, 10 m*M* KCl, 1.5 m*M* $MgCl_2$).

The SOL and CSK obtained as above were centrifuged through gradients of 20%–60% sucrose in STE (10 m*M* Tris-HCl, pH 7.4, 0.1 *M* NaCl, 5 m*M* EDTA) for 2 h at 35 000 rpm in a Hitachi RPS 40 rotor. Each fraction was analyzed by sodium dodecyl sulfate polyacrylamide gel electrophoresis (SDS-PAGE) for labeled proteins. As shown in Fig. 1, most NP and P proteins were already present in CSK at the end of the pulse, forming two major peaks across the gradient. One of them was near the bottom of the tube and the other in several top fractions. These represent the already-formed nucleocapsids with transcriptase activity and the nonassembled free forms respectively (Hamaguchi et al. 1983). Ten-minute pulse labels mostly partially completed chains of L protein (Nagai et al. 1976).

During 20-min chase, NP and P decreased in amount in the top fractions and concomitantly increased in the nearbottom fractions, being organized into the nucleocapsids. After a further chase period, virtually all the free NP and P in CSK were assembled into the nucleocapsids. Simultaneously, the L protein became clearly detectable in the nucleocapsids. Significant amounts of NP were found in the top fractions of SOL after 10-min pulse, but were rapidly chased into CSK.

These results suggest that NP and P (and probably L) become associated with CSK immediately after their synthesis and that once attached, none of these proteins leave the CSK but rather assemble there into the nucleocapsids. Essentially identical results were obtained with chick embryo (CE) and MDBK cells.

As the cells were extracted on ice, the microtubules might have been depolymerized, and the CSK obtained might have been composed mainly of microfilaments (MF) and intermediate filaments (IF). Heuser and Kirschner (1980) have shown that when such CSK are further extracted with 0.3 *M* KI, the MF are solubilized, leaving the IF almost intact. We have obtained such a KI-soluble fraction from the CSK of infected and pulse-labeled CE cells. The mature nucleocapsids in that fraction were removed by centrifugation at 100 000 *g* for 1 h and the resulting supernatant was dialyzed overnight against a physiological salt solution (0.1 *M* KCl, 2 m*M* $MgCl_2$). When the samples were centrifuged through a sucrose gradient and analyzed for each fraction using SDS-PAGE, we found that at least part of the actin molecules and other cellular proteins from uninfected cells formed a complex which sedimented near the bottom of the tube (Fig. 2, II) and that NP molecules were present coinciding exactly with the peak of that complex from the infected specimen (Fig. 2, I). Because of shutoff of host synthesis by the infection, actin molecules were clearly detected only in gels stained by Coomassie blue. Presence of ATP (0.2 m*M*) in both dialysis bags and solution promoted that coaggregation of NP and actin. These data suggest that NP and cytoskeletal proteins, possibly actin, have the capacity to interact directly with each other.

These data suggest that NDV nucleocapsid with transcriptases P and L is formed in association with CSK and that the interaction of the structural subunits, NP, with cellular CSK, probably actin filaments, may mediate this association. The initial binding of viral capsid proteins to CSK and their immobilization into a small space probably favors self- and heterointeractions, providing a basis for increased cooperative binding of viral components to form the nucleocapsid.

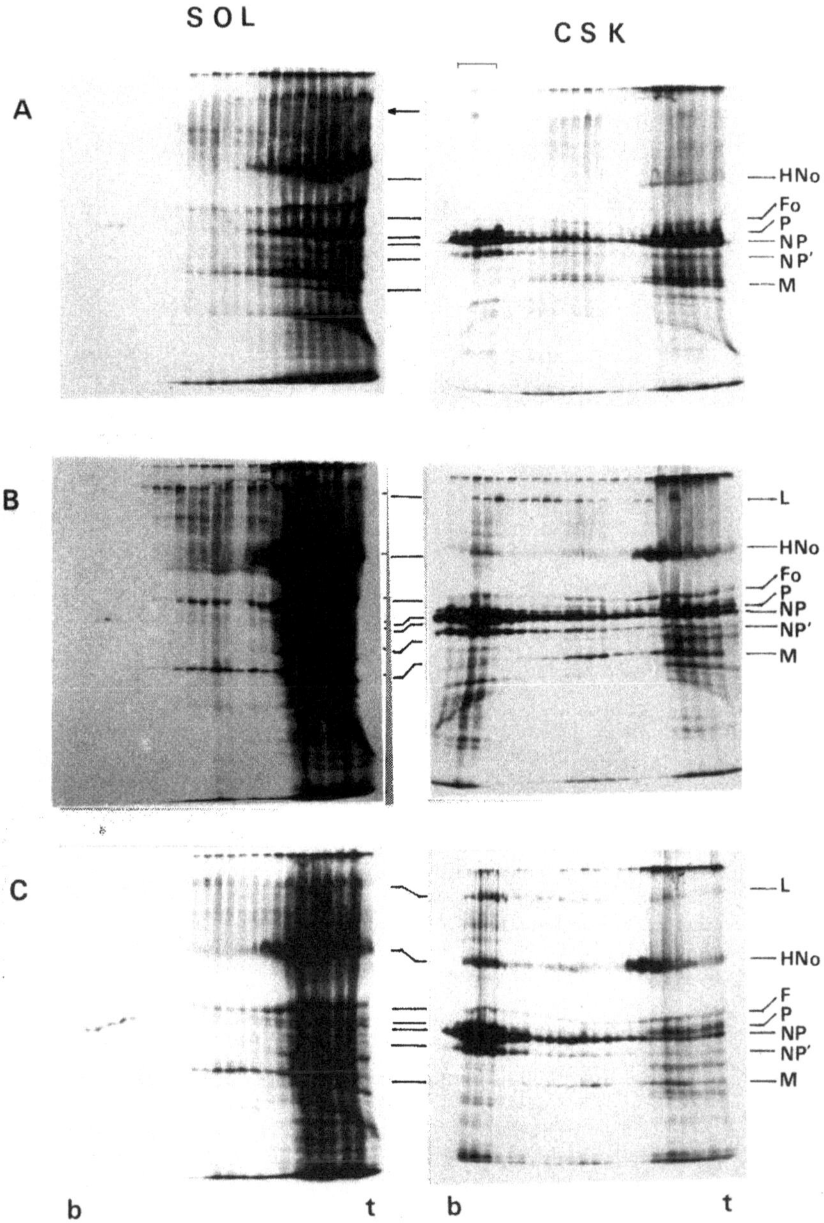


**Fig. 1 A–C.** Assembly of NDV nucleocapsids. BHK-21 cells infected with strain Ulster of NDV were labeled for 10 min with [$^{35}$S]methionine at 6.5 h after infection (**A**), or chased at that time for 20 min (**B**) and 60 min (**C**). They were fractionated into SOL and CSK, and after solubilization of CSK, both fractions were centrifuged through sucrose gradients as described by Hamaguchi et al. (1985). Fractions collected from the bottom of the tubes were analyzed using SDS-PAGE. The viral proteins organized into the nucleocapsids were recovered from near-bottom fractions (⌐¬). *t*, top; *b*, bottom

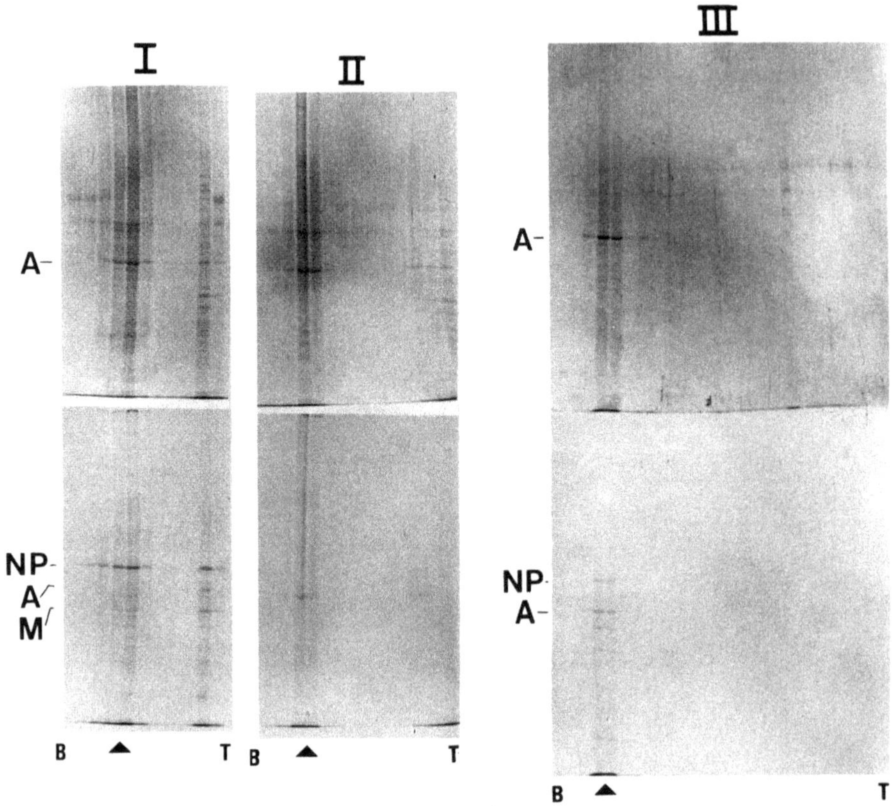


**Fig. 2.** In vitro polymerization of NP and cytoskeletal proteins. The KI-soluble fractions obtained from CSK of infected *(I, III)* and uninfected *(II)* CE cells as described in the text, were dialyzed against 0.1 *M* KCl and 2 m*M* $MgCl_2$ in the absence *(I, II)* or presence *(III)* of 0.2 m*M* ATP. They were then centrifuged as in Fig. 1, and each fraction was analyzed by SDS-PAGE. ▲, peak fractions to which actin *(A)* or actin and NP were sedimented; *B*, bottom; *T*, top; *upper gels*, stained with Coomassie blue; *lower gels*, autoradiographed

## Association of the Viral Glycoproteins with the Cytoskeletal Framework

As shown in Fig. 1, a significant portion of NDV glycoproteins, $HN_0$ and $F_0$, became associated with CSK during a 60-min chase, but not after a 10-min pulse. In contrast to our earlier observation (Hamaguchi et al. 1985), the glycoproteins were found to show marked resistance to extraction with Triton X-100 by immunofluorescent staining (not shown). We further examined the distribution of glycoproteins between SOL and CSK in combination with the use of inhibitors of their intracellular transport. As shown in Fig. 3, both carbonyl cyanide *m*-chlorophenylhydrazone (CCCP; 50 μg/ml) and monensin (1 μ*M*), which block the transport from rough endoplasmic reticulum and between Golgi stacks respectively, were found to alter the distribution markedly, so that the glycoproteins, particularly $HN_0$, were extracted predominantly into SOL throughout the chase period. Thus, the association of

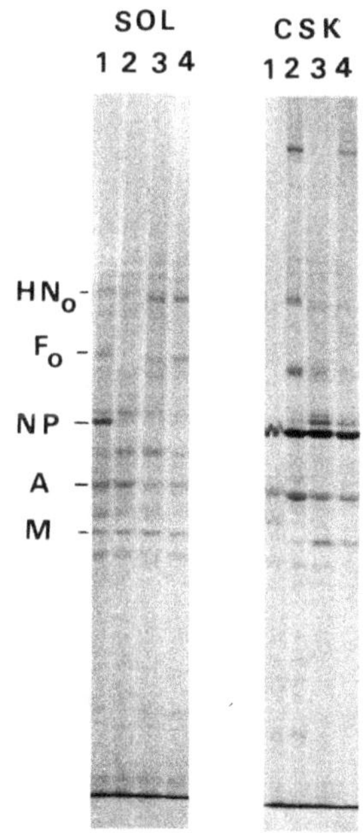


**Fig. 3.** Effect of CCCP and monensin on the distribution of NDV glycoproteins between SOL and CSK. Infected CE cells were pulse-labeled for 10 min *(1)* with [$^{35}$S]methionine and chased for 60 min without drugs *(2)* or with CCCP *(3)* and monensin *(4)*. The latter drug was present from the beginning of infection. The cells were fractionated into SOL and CSK and analyzed using SDS-PAGE

the glycoproteins with CSK could represent an event in the late replication step, possibly the virus assembly at the plasma membrane. Morphological studies of NDV (Hamaguchi et al. 1985) and measles virus (Bohn et al. 1986) infections have suggested that the glycoproteins at the budding areas are indirectly bound to CSK through the interaction of nucleocapsids with actin filaments. Further, the latter study has shown that these filaments protrude almost exclusively into the virus particles with their barbed ends. In addition, cytochalasin B, which inhibits the polymerization of barbed ends of actin filaments (McLean-Fletcher and Pollard 1980; Brown and Spudich 1981), suppresses the release of virions from cells infected with measles virus (Stallcup et al. 1983) and NDV (unpublished). Thus, it appears that forces resulting from a vectorial growth of actin filaments contribute to the formation of paramyxovirus buds. Alternatively, the cytoskeletal association of virus buds may retard the release of virus particles (Morrison and McGinnes 1985). This may explain why high amounts of progeny are cell-bound in NDV-infected cells (Nagai et al. 1976).

## Association of Viral RNA Synthesis with the Cytoskeletal Framework

The CSK and SOL fractions from CE cells infected and labeled for 60 min with [$^3$H]uridine (10 μCi/ml) were analyzed for viral RNA species contained by sucrose gradient centrifugation (Hamaguchi et al. 1985). As shown in Fig. 4, most of the

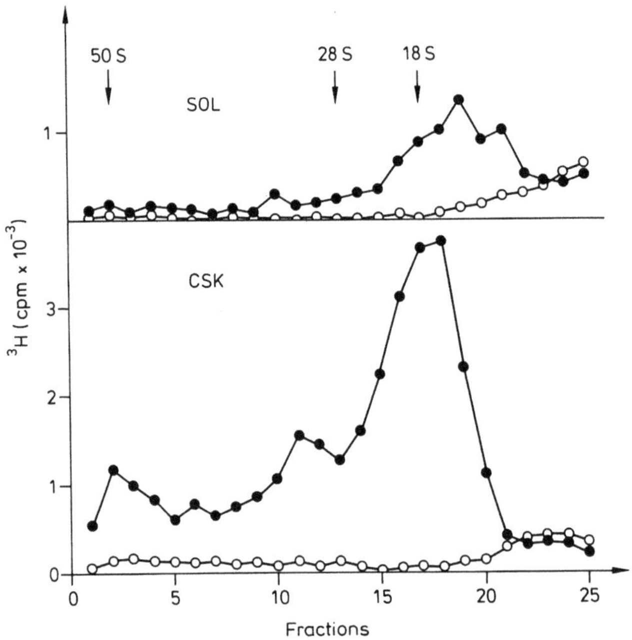


**Fig. 4.** Distribution of NDV-specific RNAs between SOL and CSK. Infected CE cells were labeled for 60 min with [$^3$H]uridine in the presence of actinomycin D, and RNAs extracted from SOL and CSK fractions were analyzed by sucrose gradient centrifugation as described by Hamaguchi et al. (1985). ●, infected; ○, mock infected

50 S RNA and the subgenomic RNAs (18–35 S) were present in CSK. On the other hand, when the distribution of total intracellular poly A RNAs was examined by means of Northern blotting using several virus-specific cDNA clones, no enrichment of RNA in CSK was observed; rather, each RNA species appeared to be more abundant in SOL (Fig. 5). Further, only those mRNAs bound to CSK appeared to participate in polysome formation, and few polysomes were present in SOL (not shown). These data, taken together, suggest that the viral mRNAs are associated with CSK when they are translated. They may thereafter leave CSK and enter SOL. A similar conclusion was reached for vesicular stomatitis virus mRNA (Cervera et al. 1981).

## Conclusions

The studies presented here indicate that virtually all of the NDV proteins sooner or later become associated with the cytoskeletal framework in infected cells. In addition, both genomic RNAs and mRNAs appear to be associated with this framework. Thus, it seems that the cytoskeletal framework might participate positively in various metabolisms essential for paramyxovirus replication. These include intermolecular

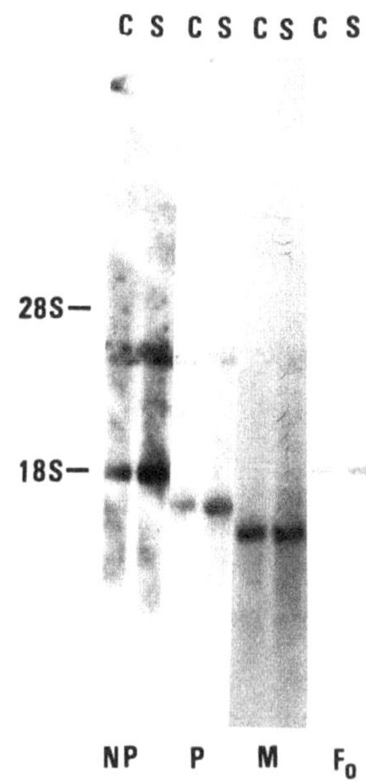


**Fig. 5.** Distribution of NDV poly A RNAs between SOL *(S)* and CSK *(C)* examined by Northern blotting. RNA extracted from SOL and CSK of infected CE cells was selected by oligo dT column chromatography and subjected to Northern blotting using cDNAs to NP, P, M and $F_0$ mRNAs. Each monocistronic mRNA is detected at or slightly below the 18 S region

interactions to form higher-order structures, such as virus nucleocapsids and virus buds, and viral genome expression. Thus, the system used here may be valid for elucidation of possible functional roles of the cytoskeleton which have not yet been recognized. It is particularly important to define the cellular and viral components and their domains which are involved in the virus-cell interactions described here.

## References

Bohn W, Rutter G, Hohenberg H, Mannweiler K, Nobis P (1986) Involvement of actin filaments in budding of measles virus: studies on cytoskeletons of infected cells. Virology 149: 91–106

Brown SS, Spudich JA (1981) Mechanism of action of cytochalasin: evidence that it binds to actin filament ends. J Cell Biol 75: 593–605

Cervera M, Breyfuss G, Penman S (1981) Messenger RNA is translated when associated with the cytoskeletal framework in normal and VSV-infected cells. Cell 23: 113–120

Hamaguchi M, Yoshida T, Nishikawa K, Naruse H, Nagai Y (1983) Transcriptive complex of Newcastle disease virus. I. Both L and P proteins are required to constitute an active complex. Virology 128: 105–117

Hamaguchi M, Nishikawa K, Toyoda T, Yoshida T, Hanaichi T, Nagai Y (1985) Transcriptive complex of Newcastle disease virus. II. Structural and functional assembly associated with the cytoskeletal framework. Virology 147: 295–308

Heuser JE, Kirschner MW (1980) Filament organization revealed in platinum replicas of freeze-dried cytoskeletons. J Cell Biol 86: 212–234

Masters C (1984) Interactions between glycolytic enzymes and components of the cytomatrix. J Cell Biol 99: 222s–225s
McLean-Fletcher S, Pollard DT (1980) Mechanism of action of cytochalasin B on actin. Cell 20: 329–341
Morrison TG, McGinnes LJ (1985) Cytochalasin D accelerates the release of Newcastle disease virus from infected cells. Virus Res 4: 93–106
Nagai Y, Ogura H, Klenk H-D (1976) Studies on the assembly of the envelope of Newcastle disease virus. Virology 69: 523–538
Penman S (1985) Virus metabolism and cellular architecture. In: Fields BN et al. (eds) Virology. Raven, New York
Stallcup KC, Raine CS, Fields BN (1983) Cytochalasin B inhibits the maturation of measles virus. Virology 124: 59–74

# Correlation of High Evolutionary Rate of Influenza A Viruses in Man with High Mutation Rate Measured in Tissue Culture: A Hypothesis

PETER PALESE

Department of Microbiology, Mount Sinai School of Medicine, One Gustave L. Levy Place, New York, NY 10029, USA

## Introduction

Influenza A viruses are intensely studied due to their association with pandemic and epidemic outbreaks in man. This disease pattern is associated with the unique ability of influenza A viruses to rapidly change both their antigenic character and their virulence traits (for reviews see Rott 1980; Scholtissek 1983; Webster et al. 1983). As soon as an immune response is generated by the host, a "new" virus appears to be selected. In this way influenza virus can effectively evade the host immune response and always find a susceptible reservoir in man. Two mechanisms have been described which contribute to this change. First, reassortment of viruses in coinfected cells may occur, leading to the introduction and expression of novel hemagglutinin (HA) genes in the virus. This event may be rare in nature, since it has led to only three new viral HA subtypes (H1, H2 and H3) in man during this century. The second type of change is mediated through the introduction of mutations in the HA genes of viruses of the same subtype, with subsequent selection of the "right" variant.

## Accumulation of Mutations in the HA Gene

Information on changes in the HA genes of influenza A virus field isolates has been generated in many laboratories. The data suggest that amino acid changes in the HA occur in areas which have been defined as antigenic sites through the use of escape mutants selected by monoclonal antibodies. Specifically, a comparison of nucleotide sequences of 19 H3 HAs from variants isolated in different parts of the world between 1968 and 1980 revealed that new H3 strains emerged via sequential amino acid changes (Both et al. 1983). These results suggest that the H3 variants evolved along a common pathway. The same result was found for H1 variants (Raymond et al. 1986). Again, the HAs of isolates showed sequential changes which suggested linear evolutionary trees. This is consistent with selection of a dominant strain derived from a preceding dominant strain.

## Linear Evolution of the NS Gene

A high level of genetic homology has been observed among NS genes of strains belonging to influenza A viruses of the H1, H2 and H3 subtypes (Scholtissek et al. 1978). This suggests that no novel NS genes were introduced through reassortment into the pool of human influenza A viruses circulating in this century. We thus chose this gene for a careful study of the evolutionary relationships among human influenza A viruses.

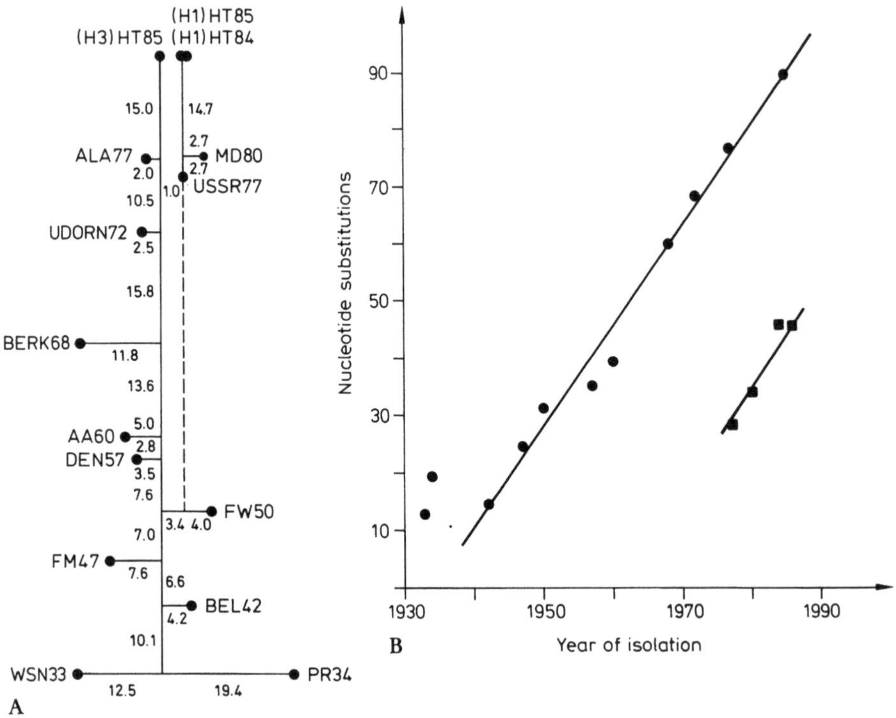


**Fig. 1 A, B.** **A** Most parsimonious evolutionary tree for 15 influenza A virus NS genes. The nucleotide sequences were analyzed using the method of Fitch (1971). The length of the trunk and side branches of the evolutionary tree are proportional to the minimal number of substitutions required to account for the difference in sequence. Nonintegral numbers arise from averaging over all possible minimal solutions. The branch defined by the USSR/77, MD/80, HT/84 and HT/85 strains is made up of isolates belonging to the H1 subtype, which was reintroduced into the human population in 1977. The *broken line* connecting this branch with the trunk of the evolutionary tree represents the predicted number of additional substitutions between the NS genes of FW/50 and USSR/77 based on the calculated evolutionary rate. **B** Linearity with time of number of substitutions in the NS genes of influenza A viruses. The *abscissa* represents the year of isolation of the influenza A viruses used in the analysis. The *ordinate* indicates the number of substitutions observed in their NS genes between the first branching point formed by the WSN/33 and PR/34 sequences in **a** and the tips of all branches of the evolutionary tree. A line generated by linear regression analysis is drawn through the points. The slope of the line is 1.7 ± 0.1 substitutions per year, which translates to 1.9 ± 0.1 substitutions per site per year. In addition to the sequences found on the trunk of the evolutionary tree *(circles)*, the NS genes of the four new H1 viruses are also represented in this graph *(squares)*. (From Buonagurio et al. 1986)

A comparison of the complete nucleotide sequences of the NS genes of 15 human influenza A viruses was carried out. The strains had been isolated over a 53-year period (1933–1985) and analysis of the NS gene sequences revealed a linear evolutionary relationship (Fig. 1 A). Further calculations involving the slope of the curve in Fig. 1 B indicate a constant rate of evolution in the NS genes of approximately $2 \times 10^3$ substitutions per nucleotide per year (Buonagurio et al. 1986). This finding was highly unusual, since it suggests strong selective forces and a high effective rate of evolution for influenza A virus genes.

The evolutionary model postulated for influenza A viruses is quite different from that suggested for influenza C and for most other viruses (Buonagurio et al. 1985). Influenza C viruses do not appear to accumulate changes rapidly with time, and different variants cocirculate at any given time. As a matter of fact, sequencing of the NS gene of an influenza A virus allows dating of the virus isolate, whereas this procedure would not help in identifying the year of isolation of an influenza C virus. Also, analysis of polio viruses isolated over the last 40 years does not suggest a constant change in the genome with time. Rather, the fact that the vaccine developed several decades ago is still in effective use suggests a conservation of the viral structure.

What could account for these great differences in evolutionary pattern among human viruses? First, we may speculate that an unusual immune response against influenza A viruses favors selection of escape variants, whereas this is not the case for other viruses. Also, fewer people may be infected at any one time by viruses such as poliovirus, insuring a larger reservoir of immunologically naive people than is the case with influenza A viruses. Thus, in order to "survive," these viruses would need no further escape mechanism. In addition to these factors, certain intrinsic properties of viruses and their replication systems may be responsible for differences in their evolutionary patterns. For example, a replication system with a high error rate may produce viral populations with great genetic variability. Furthermore, a virus whose genes (gene products) are under low functional constraint may easily permit fixation of novel mutations. Under such circumstances evolutionary rates may become high.

## Measurement of Mutation Rates of Influenza A and Poliovirus

We have attempted to elucidate the points mentioned above by directly measuring the mutation rates of RNA virus genes in tissue culture. The experimental design is shown in Fig. 2. Approximately 100 clones of influenza A/WSN and of poliovirus type 1, each of which had arisen from a single virion in one plaque generation, were obtained and amplified. One gene of each virus was chosen for sequence analysis: the NS gene of influenza A virus and the VP 1 gene of poliovirus. The latter gene had previously been shown to undergo changes and thus was known to be a gene which was not genetically conserved (Racaniello and Baltimore 1981; Nomoto et al. 1982).

Among the more than 100 000 nucleotides in each gene sequenced, seven mutations were found for the NS gene of influenza A viruses, but none for the VP 1 gene

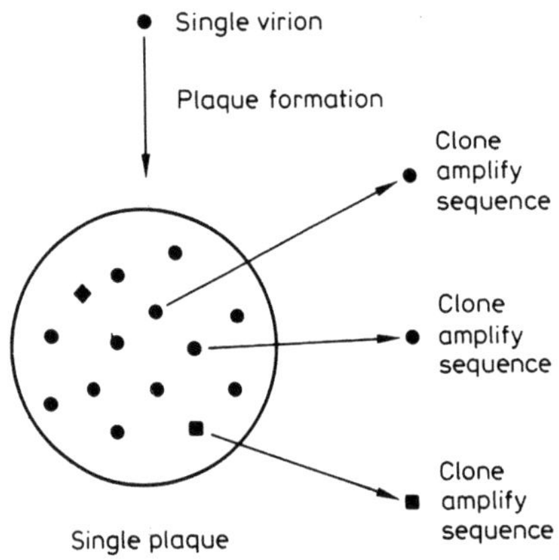


**Fig. 2.** Experimental design for measuring mutation rates in viral genes. A single virion (the parental virus) formed a plaque after sufficient time for about five infectious cycles. The many progeny virions contained within the plaque carried genes with the parental sequence *(filled circles)*, and a fraction of the virions present carried genes with a point mutation *(square* and *diamond)*. The individual virions were cloned by plaquing and amplified under conditions that minimize the effect of new mutations on the consensus sequence of the gene in the clone. RNA obtained from purified virus was sequenced using the dideoxy chain termination procedure. The mutation rate was determined by dividing the number of observed point mutations by the number of nucleotides analyzed and by the number of infectious cycles. (From Parvin et al. 1986)

of poliovirus. The calculated mutation rates are $1.5 \times 10^{-5}$ and less than $2.1 \times 10^{-6}$ –mutations per nucleotide per infectious cycle for influenza A and poliovirus respectively (Parvin et al. 1986). Thus, the mutation rate of influenza A virus appears to be significantly higher than that of poliovirus. Although different cell systems had to be used for the two viruses, and formation of defective particles, which might influence the result, could not be excluded, we believe that the experimental design allows comparison of effective mutation rates among viruses. We further speculate that although mutation rates may be high for all RNA viruses (Steinhauer and Holland 1986) there are still significant differences in mutation rates among viruses. This notion is supported by a recent study by Durbin and Stollar (1986) which shows that the RNA-containing Sindbis virus has a mutation rate lower than that of influenza A virus. Differences in mutation rates may thus explain the ease of selection for epidemiologically significant strains which survive and possess "growth" advantage in the host population.

Clearly a high mutation rate may not be the only factor contributing to rapid change in nature; rather, it may be a requisite which, together with selection and/or high transmissibility of a virus, can result in a rapid evolutionary pattern. Experiments are currently under way in our laboratory to test this hypothesis with respect to different RNA-containing viruses. Recently, unusually high variation rates have been observed for HIV isolates obtained from patients (Hahn et al. 1986). It will be interesting to compare mutation rates measured in tissue culture systems with evolutionary patterns of the respective viruses and to include RNA tumor viruses in these studies.

## References

Both GW, Sleigh MJ, Cox NJ, Kendal AP (1983) Antigenic drift in influenza virus H3 hemagglutinin from 1968–1980: multiple evolutionary pathways and sequential amino acid changes at key antigenic sites. J Virol 48: 52–60

Buonagurio DA, Nakada S, Desselberger U, Krystal M, Palese P (1985) Noncumulative sequence changes in the hemagglutinin genes of influenza C virus isolates. Virology 146: 221–232

Buonagurio DA, Nakada S, Parvin JD, Krystal M, Palese P, Fitch WM (1986) Evolution of human influenza A viruses over 50 years: rapid, uniform rate of change in NS gene. Science 232: 980–982

Durbin RK, Stollar V (1986) Sequence analysis of the E1 gene of a hyperglycosylated, host restricted mutant of Sindbis virus and estimation of mutation rate from frequency of revertants. Virology 154: 135–143

Fitch WM (1971) Toward defining the course of evolution: minimum change for a specific tree topology. Systomat Zool 20: 406–416

Hahn BH, Shaw GM, Taylor ME, Redfield RR, Markham PD, Salahuddin SZ, Wong-Staal F, Gallo RC, Parks ES, Parks WP (1986) Genetic variation in HTLV-III/LAV over time in patients with AIDS or at risk for AIDS. Science 232: 1548–1553

Nomoto A, Omata T, Toyoda H, Kuge S, Horie H, Kataoka Y, Genba Y, Nakano Y, Imura N (1982) Complete nucleotide sequence of the attenuated poliovirus Sabin 1 strain genome. Proc Natl Acad Sci USA 79: 5793–5797

Parvin JD, Moscona A, Pan WT, Leider JM, Palese P (1986) Measurement of the mutation rates of animal viruses: influenza A virus and poliovirus type 1. J Virol 59: 377–383

Racaniello VR, Baltimore D (1981) Molecular cloning of poliovirus cDNA and determination of the complete nucleotide sequence of the viral genome. Proc Natl Acad Sci USA 78: 4887–4891

Raymond RL, Caton AJ, Cox NJ, Kendal AP, Brownlee GG (1986) The antigenicity and evolution of influenza H1 hemagglutinin, from 1950–1957 and 1977–1983: two pathways from one gene. Virology 148: 275–287

Rott R (1980) Genetic determinants for infectivity and pathogenicity of influenza viruses. Philos Trans R Soc Lond [Biol] 288: 393–399

Scholtissek C (1983) Genetic relatedness of influenza viruses (RNA and protein). In: Palese P, Kingsbury DW (eds) Genetics of influenza viruses. Springer, Vienna New York, pp 99–126

Scholtissek C, Rohde W, von Hoyningen V, Rott R (1978) On the origin of the human influenza virus subtypes H2N2 and H3N2. Virology 87: 13–20

Steinhauer DA, Holland JJ (1986) Direct method for quantitation of extreme polymerase error frequencies at selected single base sites in viral RNA. J Virol 57: 219–228

Webster RG, Laver WG, Air GM (1983) Antigenic variation among type A influenza viruses. In: Palese P, Kingsbury DW (eds) Genetics of influenza viruses. Springer, Vienna New York, pp 127–168

# Cell Binding and Membrane Fusion Activities of Influenza Viruses*

J. J. SKEHEL[1] AND D. C. WILEY[2]

[1] Division of Virology, National Institute for Medical Research, Mill Hill, London NW7 1AA, UK
[2] Department of Biochemistry and Molecular Biology, Harvard University, Cambridge, MA 02138, USA

Viruses infect cells by binding to cell surface receptors and then either crossing the surface membrane directly or entering endosomes and subsequently crossing endosomal membranes. When the viruses themselves have membranes these entry processes involve membrane fusion, either between surface membrane and virus membrane or between endosomal membrane and virus membrane. The latter route of entry is used by influenza viruses, and in both the initial receptor binding function and membrane fusion activities the virus component involved is the major virus membrane glycoprotein, the haemagglutinin. The properties and structural basis of these activities is the topic of this brief review.

The haemagglutinin is a trimer of identical subunits each of which contains two glycopolypeptides, $HA_1$ and $HA_2$, and its structure is known to 3Å resolution (Wilson et al. 1981) (Fig. 1). It binds viruses to cells by interacting with membrane receptor molecules containing sialic acid. This interaction varies in detailed specificity for different influenza viruses, and comparative analysis of the haemagglutinins of viruses which demonstrate preferences for binding to carbohydrate side chains containing neuraminic acid in either $\alpha$2,6 or $\alpha$2,3 linkage with penultimate residues has been used to obtain information concerning the location of the receptor binding site on the haemagglutinin. The haemagglutinin of the Hong Kong influenza virus recognizes sialyloligosaccharides with the terminal sequence SA $\alpha$2,6 Gal-, and variants selected for their ability to grow in the presence of non-immune horse serum, a rich source of SA $\alpha$2,6 Gal-containing glycoprotein inhibitors of haemagglutinin-cell membrane interaction, have greatly increased affinity for sialyloligosaccharides with the sequence SA $\alpha$2,3 Gal-. Comparisons of the amino acid sequences of wild type and variant haemagglutinins indicate that they differ only at residue 226 of the $HA_1$ glycopolypeptide chain (Rogers et al. 1983). Mutants with different binding specificities have also been selected by growing virus in the presence of certain anti-haemagglutinin monoclonal antibodies (Daniels et al. 1987). In these cases amino acid substitutions at residues 193, 218 and a deletion from residue 225 to 230 influence receptor specificity. These observations indicate that the sialic acid binding site is a surface pocket at the distal end of the haemagglutinin molecule, the bottom of which is formed by the phenolic oxygen of tyrosine 98 and the hydrophobic surface of tryptophan 153. The back wall of the pocket has histidine 183 projecting out to

* Support is acknowledged from the Medical Research Council and from NIH AI13654 and NSF PC771398.

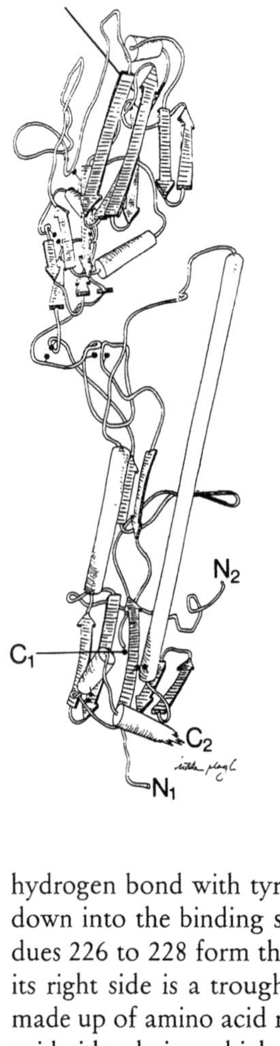


**Fig. 1.** A schematic diagram of a subunit of the Hong Kong influenza virus haemagglutinin to show the location of the sialic acid binding site *(arrow)* and the N- and C-termini of the $HA_1$ and $HA_2$ glycopolypeptide components

hydrogen bond with tyrosine 98, and both glutamic acid 190 and leucine 194 hang down into the binding site from an $\alpha$-helix which forms the top of this wall. Residues 226 to 228 form the left edge and 133 to 137 the front edge of the pocket, and its right side is a trough terminating at threonine 155. The perimeter of the site is made up of amino acid residues which have varied during antigenic drift. The amino acid side chains which form its surface and which are positioned such that they could make direct contact with the receptor include histidine 183, tyrosine 98, glutamic acid 190, and serine 136 which have exposed side chain nitrogen and oxygen atoms, and leucine 194, tryptophan 153 and leucine 226 with non-polar side chains which have the potential to make van der Waal contacts with the hydrophobic parts of the receptor. In addition the carbonyl oxygens and amide nitrogens of the peptide bonds in sections of the main chain from residues 225–228 and 131–137 have the potential to form hydrogen bonds with the receptor.

To analyse details of receptor binding we have crystallized a mutant haemagglutinin containing a glutamine at 226 and the wild-type haemagglutinin with leucine at this position and equilibrated the crystals in $\alpha$-2,3 sialyl lactose and $\alpha$-2,6 sialyl lactose respectively to form complexes. Solution of these structures should provide a complete description of the binding site and information on the basis of receptor binding specificities (Weis et al. in preparation).

Binding of virus particles to their receptors is followed by endocytosis, and the next stage in infection involves fusion of virus and endosomal membranes. In vitro, haemolysis and membrane fusion by influenza viruses have been found to be optimal at about pH 5.0, the pH of endosomes (for review see White et al. 1983). The mechanism of fusion is not known, but a number of observations indicate that at pH 5.0 haemagglutinin structure is modified and the molecule displays properties which suggest that in the process of fusion it may interact directly with the endosomal membranes. Specifically, on incubation at pH 5.0, soluble haemagglutinin released from virus particles by bromelain digestion, acquires the ability to bind to lipid vesicles, to bind detergent, or to aggregate in lipid- and detergent-free solutions. The molecules also become susceptible to proteolysis, but circular dichroism analyses indicate that the structural modifications which these observations imply do not involve gross conformational changes; they are more consistent with the relative movement of globular domains which retain their individual secondary structure (Skehel et al. 1982). The region of the haemagglutinin responsible for the pH 5.0-specific aggregation and lipid interactions appears to be the N-terminus of $HA_2$ (Fig. 1). This is the site at which the proteolytic processing of the haemagglutinin precursor necessary for fusion activation occurs. It is highly conserved and hydrophobic (Skehel and Waterfield 1975) and is released from pH 5.0 haemagglutinin aggregates by thermolysin digestion, which reverses the aggregation (Daniels et al. 1983). For these reasons it is considered that the fusion which initiates virus replication may involve the direct interaction of the $HA_2$ N-terminal region with the endosomal membrane.

In further investigations of these changes in structure and the basis of their pH dependence we made a comparative study of haemagglutinins from viruses which fuse with membranes at different pH (Daniels et al. 1985). These viruses were specifically selected for their ability to grow in the presence of amantadine hydrochloride, which is known to increase the pH of endosomes, and as a consequence their haemagglutinins undergo the characteristic pH-dependent conformational change at higher pH than wild-type virus. All of the single, or in a minority of cases double, amino acid substitutions detected in these mutant haemagglutinins appear to lower the energy barrier necessary for the conformational transition to a "fusion-active" state. One category of substitutions destabilizies the native location of the hydrophobic N-terminal region of $HA_2$, either by placing positive charges near the N-terminus or by altering the network of hydrogen bonds and non-polar contacts which stabilize this part of the molecule. The second category of substitutions distorts or destroys salt bridges and other contacts in the interfaces between the subunits of the trimer in regions of the molecule close to and at a distance from the N-terminus of $HA_2$ extending about 120 Å along the length of the trimer. Thus it appears that altering interactions between subunits of the trimer can raise the pH of fusion, which indicates either that changes in the protein conformation which uncover the hydrophobic N-terminal region of $HA_2$ require changes in the subunit interactions throughout the $HA_2$-$HA_2$ and part of the $HA_1$-$HA_1$ and $HA_2$-$HA_1$ interfaces, or that partial or total disruption of the quarternary structure of the molecule accompanies the low pH conformational change.

Our overall conclusion is that amino acid substitutions in the mutant haemagglutinins which lower the energy required to uncover the N-terminal region of $HA_2$ or to rearrange the subunit interfaces lower the concentration of protons required to trigger the conformational change. Which amino acids or salt bridges bind protons to cause the transition is unknown, but these results suggest that a number may be involved.

In agreement with the conclusion that amino acid substitutions which directly or indirectly destabilize the location of the N-terminus of $HA_2$ promote its extrusion at higher pH, we have also observed that in vitro the pH of fusion is influenced by temperature (Wharton et al. 1986). Indeed, virus-mediated membrane fusion can be triggered at neutral pH by heat, although the changes in haemagglutinin structure which accompany fusion at high temperature appear to be more extensive and less specific than those induced by incubation at the pH of fusion at normal temperature.

This, then, is a summary of the information available on the two functions of influenza haemagglutinin and the regions of the molecule involved in them. It may be that similar interactions and changes in structure will be observed in the cell-entry processes employed by other viruses and also by molecules such as toxins which must be delivered across cellular membranes.

It is also worth considering that these processes could represent targets for the design of antiviral reagents. Certainly the possibility exists of developing receptor analogues with higher affinity than natural receptors for the influenza haemagglutinin binding site which might specifically inhibit virus attachment, and considerations of the structural changes required for membrane fusion suggest that this process might also be prevented directly, as well as in the indirect manner described here involving increase of endosomal pH.

## References


Daniels RS, Douglas AR, Skehel JJ, Waterfield MD, Wilson IA, Wiley DC (1983) Studies of the influenza virus haemagglutinin in the pH 5 conformation. In: Laver WG (ed) The origins of pandemic influenza. Elsevier, New York, pp 1–7

Daniels RS, Downie JC, Hay AJ, Knossow M, Skehel JJ, Wang ML, Wiley DC (1985) Fusion mutants of the influenza virus haemagglutinin glycoprotein. Cell 40: 431

Daniels RS, Jeffries S, Yates P, Schild GC, Rogers GN, Paulson JC, Wharton SA, Douglas AR, Skehel JJ, Wiley DC (1987) The receptor-binding and membrane-fusion properties of influenza virus variants selected using anti-haemagglutinin monoclonal antibodies. EMBO J 6: 1459–1465

Rogers GN, Paulson JC, Daniels RS, Skehel JJ, Wilson IA, Wiley DC (1983) Single amino acid substitutions in influenza haemagglutinin change receptor binding specificity. Nature 304: 76

Skehel JJ, Waterfield MD (1975) Studies on the primary structure of the influenza virus haemagglutinin. Proc Natl Acad Sci USA 72: 93

Skehel JJ, Bayley PM, Brown EB, Martin SR, Waterfield MD, White JM, Wilson IA and Wiley DC (1982) Changes in the conformation of influenza virus haemagglutinin at the pH optimum of virus-mediated membrane fusion. Proc Natl Acad Sci USA 79: 968

Wharton SA, Skehel JJ, Wiley DC (1986) Studies of influenza haemagglutinin-mediated membrane fusion. Virology 149: 27

White J, Kielian M, Helenius A (1983) Membrane fusion proteins of enveloped animal viruses. Q Rev Biophys 16: 151

Wilson IA, Skehel JJ, Wiley DC (1981) Structure of the haemagglutinin membrane glycoprotein of influenza virus at 3 Å resolution. Nature 289: 366

# Molecular Characterization of H5N2 Influenza Viruses from Chickens*

ROBERT G. WEBSTER[1], WILLIAM J. BEAN[1], YOSHIHIRO KAWAOKA[1], KATHRYN DESHPANDE[1], and VICTOR A. FRIED[2]

[1] Department of Virology and Molecular Biology and [2]Department of Biochemistry, St. Jude Children's Research Hospital, 332 N. Lauderdale, P.O. Box 318, Memphis, TN 38101, USA

## Introduction

Influenza A viruses representative of each of the 13 hemagglutinin (HA) and nine neuraminidase (NA) subtypes have been isolated from aquatic birds throughout the world (Hinshaw et al. 1981). The majority of these viruses are nonpathogenic in domestic species; highly pathogenic influenza viruses are found only in the H5 and H7 subtypes. The availability of nonpathogenic and pathogenic members of the same subtype make avian influenza viruses ideal model systems for studying the molecular basis of pathogenicity. Extensive studies by Dr. R. Rott and colleagues (Rott et al. 1976, 1979, 1984) have developed the concept that pathogenicity is polygenic and depends on an optimal gene constellation. The HA is a key determinant in pathogenicity (Klenk et al. 1984). Cleavage of the HA is essential for infectivity (Klenk et al. 1975; Lazarowitz and Choppin 1975); to induce systemic infection, the HA must be cleaved in a broad range of different host cells (Rott et al. 1980).

The highly virulent H5N2 influenza A virus that occurred in chickens in Pennsylvania in October 1983 first appeared in an avirulent form in April 1983 (Bean et al. 1985). This provided a unique opportunity to study the molecular changes in a virus that occurred during the acquisition of virulence. Studies to date on this virus have concentrated on the molecular changes in the HA and NA genes (Kawaoka et al. 1984; Deshpande et al. 1985). The acquisition of virulence by A/Chicken/Pennsylvania/1370/83 [H5N2 (Chick/Penn)] has been associated with ability to plaque in tissue cultures and cleavage of HA into HA1 and HA2 in the absence of trypsin and reduction in molecular weight of HA1 commensurate with loss of a carbohydrate side chain (Kawaoka et al. 1984).

Gene reassortment studies have established that the difference between the virulent [Chick/Penn/1370/83 (vir)] and avirulent [Chick/Penn/1/83 (avir)] viruses was a mutation of the HA (Webster et al. 1986). Nucleotide sequencing analysis suggested that the critical mutation eliminated a possible site for the attachment of an *N*-linked oligosaccharide at Asn 11 of HA1 of the avirulent virus by altering a consensus sequence for glycosylation. Preliminary evidence based on molecular

* This work was supported by U.S. Public Health Research Grants AI 08831, AI 52586, AI 22275, GM 31461, and DMB83-08211 and by Cancer Center Support (CORE) Grant CA 21765, Research Development Grant RR05584-21, and American Lebanese Syrian Associated Charities.

weight differences in HA1 and growth in tunicamycin suggested loss of a carbohydrate (Kawaoka et al. 1984). However, since not all Asn-X-Ser/Thr sites are glycosylated in different influenza virus HAs, we do not know how many of the potential glycosylation sites contain carbohydrate, and since all of the influenza virus HAs studied to date contain a carbohydrate at this location in the HA molecule (Wilson et al. 1983; Keil et al. 1984), the question of glycosylation at this site remains unresolved.

To determine whether altered glycosylation of the Chick/Penn HA is sufficient to explain the differences between the avirulent and virulent strains, we have identified the sites of attachment of the *N*-linked oligosaccharides in HA1 (vir) and HA1 (avir). N-terminal sequence analysis has established that the only difference in glycosylation between HA1 (vir) and HA1 (avir) is the absence of an *N*-linked oligosaccharide at Asn 11 in HA1 (vir). This result is a direct demonstration that a site-specific glycosylation can regulate the processing of HA and thus the virulence of a virus.

The second area covered in this report is the biochemical and immunological characterization of avirulent H5N2 influenza viruses that continue to circulate in live poultry markets in New York City.

## The HA1 Protein of the Avirulent Strain Is More Extensively Glycosylated Than Is HA1 of the Virulent Strain

Chick/Penn HA1 (avir) has a slower mobility than does the HA1 (vir) when compared by SDS-PAGE. This difference, equivalent to approximately 3 kd, is illustrated in Fig. 1, showing the purified HA1 proteins of the avirulent and virulent strains in lanes 2 and 4 respectively. Digestion with *N*-glycanase eliminated this difference in mobility (Fig. 1, lanes 1 and 3), strongly suggesting that these HA1 forms differ by the number of *N*-linked oligosaccharide chains.

## The HA1 Protein of the Avirulent Strain is Glycosylated at Asn Residue 11 Whereas HA1 of the Virulent Strain Is Not

Nucleotide sequence analysis of the HA genes from the virulent and avirulent viruses suggested that the predicted amino acid change, at residue 13, eliminated a potential glycosylation site at Asn 11 on HA1 (vir) by replacing a Thr with a Lys (Kawaoka et al. 1984). However, the nucleotide sequence of this region also indicates a possible glycosylation site at Asn 10 for the HA1s of both viruses. To determine the actual glycosylation at either site, the *N*-glycanase-digested HA1 proteins were sequenced through this region, starting from the N-terminal end of the intact proteins. The analysis and observed sequence is shown in Table 1. The consequence of deglycosylation by *N*-glycanase is the conversion of the glycosylated Asn to an Asp residue at that position in the sequence (Plummer et al. 1984). The N-terminal sequences demonstrated that neither Asn 10 nor Asn 11 is glycosylated on the HA1

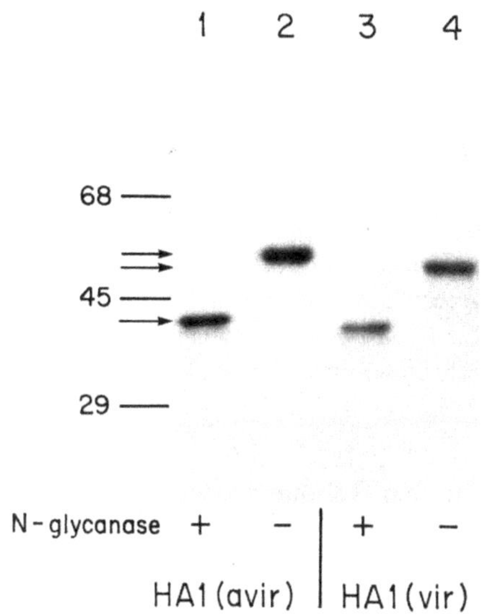


**Fig. 1.** The apparent mass difference between HA1 (vir) and HA1 (avir) is due to the content of *N*-linked oligosaccharide. HA1 (vir) and HA1 (avir) were purified by electroelution from SDS-polyacrylamide gels and aliquots digested by *N*-glycanase. Digested and control samples were analyzed on 10%–20% gradient SDS-polyacrylamide gels stained with Coomassie blue. *Lanes:* 1, HA1 (avir) + *N*-glycanase; 2, HA1 (avir) − *N*-glycanase; 3, HA1 (vir) + *N*-glycanase; 4, HA1 (vir) − *N*-glycanase. $M_r$ standards were run in adjacent lanes but are not shown

**Table 1.** Partial N-terminal sequence of deglycosylated HA1 of virulent and avirulent strains

| Cycle | HA1 (avirulent) | | HA1 (virulent) | | HA1 (revertant) | |
|---|---|---|---|---|---|---|
| | Amino acid | Yield (pmol) | Amino acid | Yield (pmol) | Amino acid | Yield (pmol) |
| 1 | D | 35 | D | 59 | D | 270 |
| 2 | Q | 17 | Q | 32 | Q | 174 |
| 3 | I | 23 | I | 47 | I | 249 |
| 4 | C | ND | C | ND | C | ND |
| 5 | I | 15 | I | 39 | I | 227 |
| 6 | G | 11 | G | 31 | G | 162 |
| 7 | Y | 14 | Y | 36 | Y | 208 |
| 8 | H | ND | H | 22 | H | 125 |
| 9 | A | 16 | A | 27 | A | 222 |
| 10 | N | 11 | N | 17 | N | 159 |
| 11 | D* | 12 | N | 21 | D* | 164 |
| | N | <0.5 | D | <0.5 | N | <1 |
| 12 | | | | | S | 29 |
| 13 | | | | | T | 67 |

The observed residues are as predicted from the nucleotide sequence, except position 11, where D (*) appears after deglycosylation of N in that position. Both N and D values are presented for residue 11 to demonstrate the point made in the text. Repetitive yield is 93%

ND, not quantitated

(vir) protein, since Asn was recovered at both positions. In contrast, Asp was recovered only at position 11 in HA1 (avir), demonstrating that this position is glycosylated. N-terminal sequence analysis of the glycosylated HA1 proteins confirmed this result, since both Asn residues were recovered in the sequence of the virulent strain whereas only a single Asn was recovered at position 10 for the avirulent strain and no residue was released at position 11 (data not shown). Glycosylated residues are not detected directly in sequence analyses. This analysis directly demonstrates that the mutation at position 13 of HA1 (vir) from Thr to Lys eliminates the functional glycosylation site at Asn 11 of HA1 (avir). Furthermore, the presence of Asn at position 10 in the HA protein of both the virulent and avirulent strains demonstrates that the sequence $Asn^{10}$-Asn-Ser is not a glycosylation site for either strain.

## The Asn Residue at Position 11 Is Glycosylated in an Avirulent Revertant of a Virulent Virus

An avirulent revertant of the virulent virus, which no longer has a cleaved HA without trypsin, has an HA1 gene sequence that differs from that of its parent by a single point mutation, resulting in a predicted amino acid change at residue 13 (Webster et al. 1986). This change, from Lys to Thr, caused reversion to a sequence containing a potential glycosylation site previously predicted for the avirulent virus (Kawaoka et al. 1984). To determine whether or not the HA of the revertant virus [HA1 (rev)] was glycosylated at Asn 11, as in the avirulent virus, the amino terminus of the HA1 (rev) was sequenced before and after *N*-glycanase treatment (Table 1). The results show that HA1 (rev), like HA1 (avir), was glycosylated at Asn 11. This suggests strongly that glycosylation at Asn 11 prevents cleavage of the A/Chicken/Pennsylvania/83 HA.

## Continuing Circulation of Avirulent H5N2 Influenza Viruses in Poultry in New York City Markets

In 1983–1984, The United States Department of Agriculture (USDA) spent approximately 61 million dollars and buried nearly 17 million birds in an attempt to eradicate the avirulent and highly virulent H5N2 influenza viruses from domestic poultry. This policy was successful at eradicating the highly virulent virus circulating in poultry in Pennsylvania and adjoining states. It is now apparent that the avirulent H5N2 virus has not been eradicated; this form of the virus is present in the live poultry markets in New York City and Miami, Florida.

The isolation of H5N2 influenza viruses from birds in live poultry markets in 1986 raises several questions: (1) Are these viruses descendents of the avirulent virus that initiated the virulent epidemic in poultry in October 1983? (2) How are these viruses maintained in city markets? (3) Do these avirulent H5N2 viruses possess a genome constellation that in combination with an appropriate hemagglutinin would be virulent?

## Are the H5N2 Viruses Circulating in City Markets in the USA Descendents of the Avirulent Viruses That Initiated the Virulent Outbreak in 1983?

Partial sequence analysis of the HA and NA genes from an H5N2 isolate from chicken (A/Ck/PA/1/86) showed that the nucleotide sequence was 98.6% and 97.9% homologous with these genes from A/Ck/PA/1370/83 [H5N2] (Table 2). Serological analysis in hemagglutination-inhibition tests with a panel of 21 different monoclonal antibodies to the HA of Ck/PA/1370/83 showed that some of the viruses are indistinguishable antigenically from the 1983 strains, while others show minor antigenic diversity (results not shown). Oligonucleotide mapping of total RNA of 25 H5N2 viruses from different species (chickens, turkeys, guinea fowl) showed that all of the strains were related to each other. Figure 2 shows the overall similarity between the H5N2 influenza viruses isolated from chickens in Pennsylvania in 1983 and from chickens in city markets in New York and Miami in 1986. When all of the oligonucleotide maps were considered, there was sufficient diversity

**Table 2.** Nucleotide sequence homology between H5N2 influenza viruses isolated in 1983[a] and 1986[b]

| Gene | Percent sequenced | No. changes/ no. sequenced | Percent homology |
|---|---|---|---|
| HA | 50 | 11/809 | 98.6 |
| NA | 32 | 9/438 | 97.9 |

[a] A/Ck/PA/1370/83
[b] A/Ck/PA/1/86

Ck/PN/29708/84 Ck/PN/2/86 Ck/FL/27715-3/86

**Fig. 2.** Oligonucleotide mapping analysis of the RNAs of avirulent H5N2 isolates. RNAs were digested with RNase T1 and P32-labeled, and the oligonucleotides were separated as described (Bean et al. 1985). The *arrows* in the center panel indicate unique oligonucleotides common to all three isolates

to separate them into four homogeneous groups indicating that all of these viruses have been circulating in poultry for some time and are showing distinguishable lines of evolution.

## How Are the H5N2 Viruses Maintained in New York City Markets?

There are over 40 live poultry markets in New York City that handle an excess of 50,000 birds per week. The range of poultry in a market includes chickens, turkeys, guinea fowl, ducks, geese, pigeons, rabbits, guinea pigs, and sometimes larger animals such as sheep. The animals are kept under crowded conditions, sometimes with different species in one cage. The poultry come from surrounding states and the majority pass through a central supply market: extensive surveillance indicated that influenza viruses were not being regularly introduced into the city markets from the large commercial suppliers. In the city markets, chickens turn over within 1–5 days while guinea fowl and turkeys may stay for up to 7 days or more.

Detectable levels of H5N2 virus were present in birds from the majority of the New York markets in February and March 1986. Influenza viruses of the H5N2 subtype isolated repeatedly from the same market during the period February-April 1986 were analyzed by oligonucleotide mapping and with monoclonal antibodies to the HA. These analyses indicated that the same virus was retained in an individual market. There was presumably sufficient turnover of susceptible birds to maintain the viruses in the markets. In order to break the virus cycle within a single market, all birds were sold and extensive cleaning and disinfection was successful in markedly reducing the number of markets with H5N2 influenza virus infections. Since the same crates were originally used to transport different species without cleaning, it is possible that the outbreaks of disease on farms in Pennsylvania may have originated from the markets. All crates are now washed at the central market and despite extensive virus and serological studies, there is no evidence for reintroduction of H5N2 influenza viruses into the markets from the suppliers, indicating that the viruses are maintained in the poultry in the markets.

## Pathogenicity of Reassortant H5N2 Viruses

The H5N2 viruses isolated from birds in the live poultry markets cause no disease signs in poultry, replicate in the respiratory tract, and are not usually shed in the feces. It could be argued that the H5N2 influenza viruses from the New York City markets may have changed sufficiently so that they are no longer a significant threat to domestic poultry. In order to provide some information on this question, reassortant influenza viruses were prepared containing the HA gene from the Ck/PA/1370/83 highly pathogenic strain and other genes from H5N2 viruses isolated from chickens in 1986. These reassortants were made to retain only the HA gene from the virulent 1983 parent and as many genes as possible from the 1986 H5N2 viruses. The virus reassortants were highly pathogenic for chickens showing all the disease signs of the highly pathogenic H5N2 strain from 1983. Five different H5N2 viruses from the New York City markets have been used to make reassortants containing

Ck/PA/1370/83 HA; these have been tested in chickens and three are highly pathogenic. These are preliminary results, for each of the genes present in the reassortants has not yet been identified.

These preliminary results reemphasize the role of the HA gene in determining pathogenicity and suggests that many of the H5N2 viruses circulating in poultry in the New York markets have the gene constellation required to compliment the HA in producing lethal disease in poultry.

## Discussion

We have determined the sites at which *N*-linked oligosaccharides are attached to the HA1 subunit of the avirulent and virulent A/Chicken/Pennsylvania/83 (H5N2) virus and provide evidence that site-specific glycosylation at Asn 11 can influence cleavage of the Chick/Penn HA in tissue culture and thus the acquisition of virulence in vivo. Virulence is considered to be a polygenic property (Scholtissek et al. 1977; Ogawa and Ueda 1981; Rott et al. 1976, 1979, 1984). However, reassortant studies have shown that the avirulent Chick/Penn virus contains seven of the eight genes necessary for virulence, requiring only mutation of the HA gene for the acquisition of virulence (Webster et al. 1986). The study confirms the predictions of Kawaoka et al. (1984) and Webster et al. (1986) indicating a relationship between glycosylation of the Chick/Penn HA and the acquisition of virulence.

Through direct amino acid sequence analysis, we have shown that the HA1 subunits of the avirulent and avirulent revertant viruses differ from that of the virulent virus by containing a carbohydrate at Asn 11. An Asn residue can be glycosylated if it is part of the sequence Asn-X-Thr/Ser (Clamp and Johnson 1972); however, glycosylation did not occur on any of the three Chick/Penn HA1 subunits at the first potential site, Asn 10, while it did take place at all four remaining sites.

There are precedents for lack of glycosylation among influenza virus HAs. The HA1 subunits of A/Japan/57 (H2) and A/FPV/Rostock/34 (H7) are also not glycosylated at all potential sites (Keil et al. 1979, 1984). Absence of glycosylation at the predicted sites in these two viruses has been attributed either to the placement of an Asp or a Pro in the variable position of the glycosylation sequence (Ronin et al. 1981; Struck and Lennarz 1979), or to steric hindrance when the potential sites were adjacent (Keil et al. 1984). Neither explanation applies to the HA1 of the virulent Chick/Penn virus, which contains the sequence $Asn^{10}$-Asn-Ser-Lys.

It has been proposed that cleavage of the FPV HA depends on the presence of several consecutive basic amino acids at the cleavage site in the connecting peptide region between HA1 and HA2 (Bosch et al. 1981; Garten et al. 1982). The deduced amino acid sequence of the putative connecting peptide is the same for the virulent, avirulent, and revertant Chick/Penn HA and is similar to that of the FPV HA in that it contains at least four basic amino acids at the predicted C-terminus of HA1 (Kawaoka et al. 1984). Only influenza viruses of the H5 and H7 subtypes have been found to be virulent for avian species, and it is possible that the type of connecting peptide sequence represented by FPV is important in determining HA cleavage in infected birds. Thus, it is likely that both the sequence of the putative connecting peptide and glycosylation at Asn 11 is important in determining cleavage of the Chick/Penn HA.

It is still unresolved how glycosylation affects cleavage of the A/Chicken/Pennsylvania/83 HA. The presence of carbohydrate may mask a recognition signal required by the cleavage enzyme or perturb a structural feature so that it no longer serves as a substrate for the processing enzyme. Alternatively, the carbohydrate may be involved in generating the correct signal for trafficking the HA through the Golgi to encounter the cleavage enzyme and/or for proper assembly of cleaved HA onto virions. It is also possible that the presence of a basic residue, the Lys at position 13 in HA1 (vir), plays an additional role by serving as part of the recognition signal for the processing enzyme. Studies are in progress to define the molecular events involved in processing of the HA to HA1 and HA2.

The avirulent H5N2 influenza viruses isolated from birds in the live poultry markets in New York City from January to April 1986 may have originated from Pennsylvania in 1983 or may be the source of the viruses for chickens in Pennsylvania in April 1983. Based on the mutation rate in the HA gene and the diversity of HA sequences in 1983–1984 (Kawaoka et al. 1984; Kawaoka and Webster 1985), it can be predicted that the H5N2 influenza virus was probably in poultry 6 months before it was detected in April 1983.

## Conclusions

We have identified the glycosylation sites on the HA1 subunit from the virulent and avirulent strains of A/Chick/Penn/83 (H5N2) influenza viruses by direct structural analysis of the isolated proteins. Our results show that the only difference in glycosylation between the HA1 of the virulent and avirulent strains was the lack of an Asn-linked carbohydrate on the virulent HA1 polypeptide at residue 11. Further, we show that the HA1 of both the avirulent and virulent viruses is not glycosylated at one potential site, while all other sites contain carbohydrate. Amino acid sequence analysis of the HA1 of an avirulent revertant of the virulent strain confirmed these findings.

Based on sequence analysis of the HA and NA genes, oligonucleotide mapping of the RNA, and analysis of the HAs with monoclonal antibodies, it was shown that the viruses circulating in birds in city markets in the USA during 1986 are descendants of the H5N2 influenza viruses that appeared in chickens in Pennsylvania.

*Acknowledgments.* Lisa Newberry, John Kayoma, Hosea Clariette, and Kristi Weiner provided excellent technical assistance. We thank Lisa Wilson for typing the manuscript.

## References

Bean WJ, Kawaoka Y, Wood JM, Pearson JE, Webster RG (1985) Characterization of virulent and avirulent A/Chicken/Pennsylvania/83 influenza A viruses: potential role of defective interfering RNAs in nature. J Virol 54: 151–160

Bosch FX, Garten W, Klenk HD, Rott R (1981) Proteolytic cleavage of influenza virus hemagglutinins: primary structure of the connecting peptide between HA1 and HA2 determines proteolytic cleavability and pathogenicity of avian influenza viruses. Virology 113: 725–735

Clamp JR, Johnson I (1972) Glycoproteins, 2nd edn. Elsevier, Amsterdam

Deshpande KL, Naeve CW, Webster RG (1985) The neuraminidases of the virulent and avirulent A/Chicken/Pennsylvania/83 (H5N2) influenza A viruses: sequence and antigenic analysis. Virology 147: 49–60

Garten W, Linder D, Rott R, Klenk HD (1982) The cleavage site of the hemagglutinin of fowl plague virus. Virology 122: 186–190

Hinshaw VS, Webster RG, Rodriguez RJ (1981) Influenza A viruses: combinations of hemagglutinin and neuraminidase subtypes isolated from animals and other sources. Arch Virol 67: 191–201

Kawaoka Y, Naeve CW, Webster RG (1984) Is virulence of H5N2 influenza viruses in chickens associated with loss of carbohydrate from the hemagglutinin? Virology 139: 303–316

Kawaoka Y, Webster RG (1985) Evolution of the A/Chicken/Pennsylvania/83 (H5N2) influenza virus. Virology 146: 130–137

Keil W, Klenk HD, Schwartz RT (1979) Carbohydrates of influenza virus. III. Nature of the oligosaccharide-protein linkage in viral glycoproteins. J Virol 32: 253–256

Keil W, Niemann H, Schwartz RT, Klenk HD (1984) Carbohydrates of influenza virus. V. Oligosaccharides attached to individual glycosylation sites of the hemagglutinin of fowl plague virus. Virology 133: 77–91

Klenk HD, Rott R, Orlich M, Blödorn J (1975) Activation of influenza A viruses by trypsin treatment. Virology 68: 426–439

Klenk HD, Garten W, Bosch FX, Rott R (1984) The role of the hemagglutinin as a determinant for the pathogenicity of avian influenza viruses. In: Stuart-Harris CH, Potter CW (eds) The molecular virology and epidemiology of influenza. Academic Press, New York

Lazarowitz SG, Choppin PW (1975) Enhancement of the infectivity of influenza A and B viruses by proteolytic cleavage of the hemagglutinin polypeptide. Virology 68: 440–454

Ogawa T, Ueda M (1981) Genes involved in the virulence of an avian influenza virus. Virology 113: 304–313

Plummer TH Jr, Elder JH, Alexander S, Phelan AW, Tarentino AL (1984) Demonstration of peptide: N glycosidase F activity in Endo-$\beta$-N-acetylglucosaminidase F preparations. J Biol Chem 259: 10700–10704

Ronin C, Granier C, Caseti C, Bouchilloux S, van Rietschoten J (1981) Synthetic substrates for thyroid oligosaccharide transferase. Eur J Biochem 118: 159–164

Rott R, Orlich M, Scholtissek C (1976) Attenuation of pathogenicity of fowl plague virus by recombination with other influenza A viruses nonpathogenic for fowl plague virus. Nonexclusive dependence of pathogenicity on hemagglutinin and neuraminidase of the virus. J Virol 19: 54–60

Rott R, Orlich M, Scholtissek C (1979) Correlation of pathogenicity and gene constellation of influenza viruses. III. Non-pathogenic recombinants derived from highly pathogenic parent strains. J Gen Virol 44: 471–477

Rott R, Reinacher M, Orlich M, Klenk HD (1980) Cleavability of hemagglutinin determines spread of avian influenza viruses in the chorioallantoic membrane of chicken embryo. Arch Virol 65: 123–133

Rott R, Scholtissek C, Klenk HD (1984) Alterations in pathogenicity of influenza virus through reassortment. Lerner R, Chanock R (eds) In: Modern approaches to vaccines. Cold Spring Harbor Press, Cold Spring Harbor, pp 345–356

Scholtissek C, Rott R, Orlich M, Harms E, Rohde W (1977) Correlation of pathogenicity and gene constellation of an influenza A virus (fowl plague). 1. Exchange of a single gene. Virology 81: 74–80

Struck DK, Lennarz WJ (1979) The biochemistry of glycoproteins and proteoglycans. Plenum Press, New York

Webster RG, Kawaoka Y, Bean WJ (1986) Molecular changes in A/Chicken/Pennsylvania/83 (H5N2) influenza virus associated with acquisition of virulence. Virology 149: 165–173

Wilson IA, Ladner RC, Skehel JJ, Wiley DC (1983) The structure and role of the carbohydrate moieties of influenza virus hemagglutinin. Biochem Soc Trans 11: 145–147

# Viruses, Amyloidosis, and Dementia: Prospects for Unconventional Slow Infections

HEINO DIRINGER, MARKUS CZUB, HENK RONALD BRAIG, WERNER IMRECKE, and HARTMUT BLODE

Robert Koch-Institut des Bundesgesundheitsamtes, Nordufer 20, 1000 Berlin 65, Federal Republic of Germany

## Amyloidosis

In 1854 Rudolf Virchow published a manuscript whose title can be translated "A substance with the chemical reactivity of cellulose in human brain and spinal cord." (Virchow 1854 a). At the same time he suspected that this substance or these "corpora amylacea," which he happened to discover by staining brain sections with iodine and dilute sulfuric acid, could have developed during a pathological process. Virchow was right with his latter assumption but wrong with his idea of having discovered cellulose in the human central nervous system (CNS). He actually stained amyloid deposits in brain, the major constituent of which is now known to be protein. Virchow also recognized the similarity of the stained material in brain and deposits found in "waxy" livers and spleens (Virchow 1854 b). Since then amyloid has been found to be associated with a variety of diseases and the short introduction on amyloidosis given in our article refers to two excellent reviews of the field (Cohen 1967; Glenner 1980). In amyloidosis or $\beta$-fibrillosis, peptides with a high degree of $\beta$-pleated sheets aggregate and form peculiar ($\beta$-pleated) fibrils, insoluble under physiological conditions. These become deposited intracellularly as fibrils or extracellularly as fibril-containing amyloid plaques. The actual reason for the formation of such fibrils is unknown. In some instances, however, disturbed catabolism of proteins seems to be involved, as some amyloid peptides isolated from the deposits have originated from longer, well characterized, and sequenced precursor proteins.

Amyloid deposits are recognized and diagnosed by histological staining techniques.

Amyloidoses have been classified as systemic or organ-specific depending on the distribution of the deposits, which may involve several organs or may be restricted to just one organ. The terms "primary amyloidosis" and "secondary amyloidosis" have been used, according to whether or not another disease can be associated with the amyloidosis. Primary amyloidoses are often, but not necessarily, organ-specific and are suspected to be involved in processes of aging. Secondary amyloidosis can be induced by tumor growth, immune disturbance, metabolic diseases, and by bacterial infections such as tuberculosis or leprosy.

In this communication, we will show that viruses can also induce amyloidosis. The viruses concerned are the unconventional ones causing non-inflammatory, always fatal diseases of the CNS in men and animals. In man, they can cause Creutzfeldt-Jakob disease (CJD), a form of dementia. We will show that they most likely induce an organ-specific amyloidosis, i.e., an amyloidosis of the CNS, and – what may turn out to be interesting – an amyloidosis that cannot necessarily be recognized by histological examination techniques.

## Unconventional Slow Viruses

Unconventional infectious agents (for review see Gajdusek 1985) are, by size, to be classified as viruses. They have unconventional properties, for example the apparent absence of any immune reaction of the infected host. The term "slow" refers to the development of a disease that in the natural hosts is characterized by a preclinical phase of several months to 2 years.

These viruses replicate in the spleen and lymph nodes before entering the CNS via the blood stream. Replication in the periphery and transport via the bloodstream is characteristic of most conventional virus infections of the CNS.

Model systems have been developed. Of particular interest are those of scrapie in mice and hamsters (Kimberlin 1976; Kimberlin and Walker 1977), the latter being used extensively for virological and biochemical studies, as for example in the experiments described here.

At the moment, to our knowledge nobody has ever characterized an unconventional virus by electron microscopy, nor has anybody ever characterized a proteinaceous component or a nucleic acid as a constituent of such a virus. Nevertheless, a recent individual theory was proposed based on two major postulates: (1) the size of the infectious unit may be smaller than that of a virus (Prusiner 1982); (2) a protein component of the infectious agent has been characterized (Prusiner 1985). Postulate 1 was soon disproved (Diringer and Kimberlin 1983), and postulate 2 seems rather unlikely on the basis of experiments reported here and recently (Braig and Diringer 1985; Czub et al. 1986a, b).

## Scrapie-Associated Fibrils and Their Relation to Infectivity

Since the development of the scrapie hamster model the major advance in unconventional virus research has been the discovery of scrapie-associated fibrils (SAF) (Merz et al. 1981), now known to be specific for the whole group of unconventional slow virus diseases (Merz et al. 1983, 1984), and the discovery of a disease-specific protein (Bolton et al. 1982) now recognized as the major component of SAF, i.e., SAF protein (Diringer et al. 1983), also known as prion protein (Prusiner et al. 1983).

The copurification of infectivity with SAF and the increased specific activity in fractions containing high amounts of fibrils raised the possibility that SAF was a form of the infectious virus (Diringer et al. 1983; Merz et al. 1983; Prusiner et al. 1983). However, from the time of their discovery it was also apparent that these fibrils could be a pathological product, in this case, on grounds of morphologic appearance (Merz et al. 1981), staining properties (Prusiner et al. 1983), chemical analysis (Multhaup et al. 1985) and cloning and sequencing data (Chesebro et al. 1985; Oesch et al. 1985), an amyloid. In vivo experiments now reveal that SAF are indeed a virus-induced pathological protein and can be separated from infectivity in a kinetic study (Braig and Diringer 1985; Czub et al. 1986a). Such an experiment was performed in the following way (Czub et al. 1986a): A group of hamsters was infected intraperitoneally with scrapie virus. Subsequently at various times animals were killed and used individually to assay the content of infectivity as well as of SAF in their brains. Interesting results emerged.

The first animals became sick at 95 days post infection, marking the beginning of the clinical phase. Important things had already happened, however, in the preclinical phase. Infectivity could be detected relatively early, at 10 days post infection, but apparently remained constant up to about 50 days post infection. This had already been observed in an earlier experiment and had been interpreted as evidence for the hematogenous spread of virus to the CNS (Diringer 1984). Soon after 50 days a surprisingly active phase of virus replication occurs. This phase ceases at about 70 days having already reached maximum levels of infectivity in brain, although the phase of clinical disease is still far off.

SAF cannot be detected during the early silent phase or the phase of rapid virus replication. Minute amounts of SAF appear in the final phase of virus production, increase for about 3–4 logs in the subsequent phase, and have reached almost maximum levels at the time when clinical symptoms begin. Obviously the kinetic analysis tells us that virus replication and the formation of SAF are consecutive processes, and therefore that the fibril structure or the fibril protein is unlikely to be essential for infectivity.

Is there another possibility to test this conclusion? We have performed an experiment with spleens of mice and hamsters (Czub et al. 1986b) known to contain $10^6$ infectious scrapie units per gram of tissue. This equals about the infectivity present in 1/100–1/1000 of a hamster brain. As we have learned from our kinetic experiment, after biochemical enrichment of SAF (Multhaup et al. 1985), we can detect the SAF protein in 1/5000 equivalent of hamster brain in immunoblots (Bode et al. 1985). Thus we should be able to find SAF protein easily in preparations obtained from 2 g spleen, if SAF and infectivity correlate with each other. The search for SAF in 2 g spleen material by this method gave completely negative results. SAF protein was not detected. If, however, as a control, before biochemical purification of SAF, we added 1/100 of a scrapie brain equivalent to 2 g spleen material, we recovered SAF protein without any difficulties. This experiment, therefore, supports the conclusion we drew from the kinetic experiment: scrapie virus and scrapie-associated fibrils are different entities.

## Consequences and Future Research Prospects

### Unconventional Viruses

What are the consequences of our experiments and what future research prospects can be discerned?

First of all, for the practitioner the termination of virus replication before clinical symptoms suggests that virostatics are not indicated for treatment of patients with clinical symptoms. In addition, early life-saving diagnosis based on an immunoassay of SAF protein also seems questionable, since SAF are a consequence of virus replication.

The kinetics of virus replication as a whole, i.e., a long phase of silent presence of virus, a subsequent relatively short period of active replication, and a final long period of unchanged high virus titres, perhaps indicates a kind of "one-step growth" for scrapie virus in brain. The only other explanation for our observations seems to be that after replication is initiated, infectivity spreads from cell to cell during the active phase of replication in a form different from the completed virus. If the hypothesis of one-step growth were correct, the low amount of virus present initially would support the idea that only a very limited number of replication sites in the central nervous system are available for efficient virus replication (Dickinson and Outram 1979). It also would show why all the many trials to establish reasonable tissue culture systems for unconventional viruses have failed.

Plateau levels of infectivity before clinical signs were described two decades ago (Eklund et al. 1967), and in 1982 the authors of a histopathological time-course study on brains of spider monkeys infected intracerebrally with kuru agent came to the conclusion that not the virus is slow, but the build-up of a number of affected cells sufficiently large to cause neurological signs (Beck et al. 1982).

We now can start again to search for the virus in brain, this time at a moment after infection when large amounts of SAF have not yet built up. Hopefully, we will eventually pin down protein and nucleic acid as proper constituents of a proper virus-sized infectious agent.

## Scrapie-Associated Fibrils

SAF and SAF protein will remain leading markers in research with unconventional viruses, although, or because, they are not a component of the virus. Why? The answer is that they will enable us to study the biochemical processes involved in the formation of an amyloid deposited in the central nervous system. The information of SAF protein has been shown to be encoded in the genetic make-up of the healthy host (Oesch et al. 1985; Chesebro et al. 1985). According to one of the investigations (Chesebro et al. 1985) it seems likely that this information is transcribed into mRNA and thus translated into protein in brain only.

If this turns out to be correct, the restriction of histopathological lesions can easily be explained. The SAF protein is derived from a precursor protein of bigger size, the function of which is unknown. SAF protein is actually a glycoprotein (Manuelidis et al. 1985; Multhaup et al. 1985) and seems to be associated with membranes according to purification protocols. Its genetic information is conserved (Westaway and Prusiner 1986) during species development. Antibodies directed toward SAF protein from hamsters are known to react with that of mice (Bode et al. 1985) and of patients who have died of CJD (Bendheim et al. 1984; Bode et al. 1985; Manuelidis et al. 1985).

Furthermore, it is known that the precursor, in contrast to SAF protein, is sensitive to protease treatment (Oesch et al. 1985) and cannot be sedimented by high-speed centrifugation (Blode and Diringer, unpublished). The concentration of the precursor in brain does not change during the disease and is in the same order as that of SAF protein in scrapie brains, i.e., about 10–20 μg per brain (Blode et al., unpublished). Thus it should be of interest to resolve the biological function of the precursor and to find out the molecular events that lead to the formation of SAF as a consequence of virus replication in perhaps a limited number of cells in the CNS.

## Degenerative Diseases of the CNS and Dementia

As mentioned before, SAF and thus SAF protein can be isolated from brains of patients who have died of CJD, a transmissible dementia. If we accept that with SAF we deal with an amyloid [and recent staining experiments strengthen this view (Kimberlin 1986)], and if the reader remembers what we said earlier, that this amyloid is not necessarily detectable in histological sections of brains affected by unconventional viruses, then it seems permissible to argue that amyloidoses of the CNS are somtimes only recognized after biochemical isolation of the amyloid, or – now extending our argument more speculatively – that some degenerative diseases and dementias can be diagnosed as an amyloidosis only by biochemical techniques.

In other words, we postulate that degenerative diseases and dementias of this type in general are amyloidoses and that it might be worthwhile to look for amyloids biochemically in other uncharacterized degenerative diseases in the future. Analyses by electron microscopy have been performed already, and a table summarizing these results (Iqbal and Wisniewski 1983) supports the proposed hypothesis.

Characterization of the amyloids and their precursors may, at some time in the future, enable us to classify degenerative diseases and dementias in terms of the generated amyloid.

Keeping this in mind, scrapie research today seems to resemble tumor virus research a few decades ago, when it became obvious that a variety of noxae may force normal cells to escape regulated growth and to develop into tumor cells. Tumor virus systems have been used efficiently to gain insight into the biochemistry of this transformation. In analogy, a normal brain could be converted into a degenerated brain by a variety of toxins causing brain-specific amyloidosis. Unconventional virus systems may turn out to be good tools to help us learn something about the biochemical processes leading to degeneration of the brain and to dementia (Diringer 1985a, b).

## References


Beck E, Daniel PM, Davey AJ, Gajdusek DC, Gibbs Jr CJ (1982) The pathogenesis of transmissible spongiform encephalopathy: an ultrastructural study. Brain 105: 755–786

Bendheim PE, Barry RA, DeArmond SJ, Stites DP, Prusiner SB (1984) Antibodies to a scrapie prion protein. Nature 310: 418–421

Bode L, Pocchiari M, Gelderblom H, Diringer H (1985) Characterization of anitsera against scrapie-associated fibrils (SAF) from affected hamsters and cross-reactivity with SAF from scrapie-affected mice and from patients with Creutzfeldt-Jakob disease. J Gen Virol 66: 2471–2478

Bolton DC, McKinley MP, Prusiner SB (1982) Identification of a protein that purifies with the scrapie prion. Science 218: 1309–1311

Braig HR, Diringer H (1985) Scrapie: concept of a virus induced amyloidosis of the brain. EMBO J 4: 2309–2312

Chesebro B, Race R, Wehrly K, Nishio J, Bloom M, Lechner D, Bergstrom S, Robbins K, Mayer L, Keith JM, Garon C, Haase A (1985) Identification of scrapie prion-protein-specific mRNA in scrapie-infected and uninfected brain. Nature 315: 331–333
Cohen AS (1967) Amyloidosis. N Engl J Med 277: 522–530, 574–583, 628–638
Czub M, Braig HR, Diringer H (1986a) Pathogenesis of scrapie: study of the temporal development of clinical symptoms, of infectivity titres and scrapie-associated fibrils in brains of hamsters infected intraperitoneally. J Gen Virol 67: 2005–2009
Czub M, Braig HR, Diringer H (1986b) The major protein of SAF is absent from spleen and thus not an essential part of the scrapie agent. Arch Virol 91: 383–386
Dickinson AG, Outram GW (1979) The scrapie replication-site hypothesis and its implications for pathogenesis. In: SB Prusiner and WJ Hadlow (eds) Slow transmissible diseases of the nervous system, Academic, New York vol 2, pp 13–31
Diringer H (1984) Sustained viremia in experimental hamster scrapie. Arch Virol 82: 105–109
Diringer H (1985a) Unconventional slow viruses, amyloidosis, and dementias. Lancet ii: 661–662
Diringer H (1985b) Ein Modell für Alterungsprozesse und degenerative Veränderungen des Gehirns – Scrapie. Funkt Biol Med 4: 129–140
Diringer H, Kimberlin RH (1983) Infectious scrapie agent is apparently not as small as recent claims suggest. Biosci Rep 3: 563–568
Diringer H, Gelderblom H, Hilmert H, Özel M, Edelbluth C, Kimberlin RH (1983) Scrapie infectivity, fibrils and low molecular weight protein. Nature 306: 476–478
Eklund CM, Kennedy RC, Hadlow WJ (1967) Pathogenesis of scrapie virus infection in the mouse. J Infect Dis 117: 15–22
Gajdusek DC (1985) Unconventional viruses causing subacute spongiform encephalopathies. In: Fields BN et al. (eds) Virology. Raven, New York pp 1519–1557
Glenner GG (1980) Amyloid deposits and amyloidosis. N Engl J Med 302: 1283–1292, 1333–1343
Iqbal K, Wisniewski HM (1983) Neurofibrillary tangles. In: Reisberg B (ed) Alzheimer's disease. Free Press, New York pp 48–56
Kimberlin RH (1976) Experimental scrapie in the mouse: a review of an important model disease. Sci Prog 63: 461–481
Kimberlin RH (1986) Scrapie. How much do we really understand? Neuropathol Appl Neurobiol 12: 131–147
Kimberlin RH, Walker CA (1977) Characteristics of a short incubation model of scrapie in the golden hamster. J Gen Virol 34: 295–304
Manuelidis L, Valley S, Manuelidis EE (1985) Specific proteins associated with Creutzfeldt-Jakob disease and scrapie share antigenic and carbohydrate determinants. Proc Natl Acad Sci USA 82: 4263–4267
Merz PA, Somerville RA, Wisniewski HM, Iqbal K (1981) Abnormal fibrils from scrapie-infected brain. Acta Neuropathol (Berl) 54: 63–74
Merz PA, Somerville RA, Wisniewski HM, Manuelidis L, Manuelidis EE (1983) Scrapie associated fibrils in Creutzfeldt-Jakob disease. Nature 306: 474–476
Merz PA, Rohwer RG, Kascsak R, Wisniewski HM, Somerville RA, Gibbs CJ Jr, Gajdusek DC (1984) Infection specific particle from the unconventional slow virus diseases. Science 225: 437–440
Multhaup G, Diringer H, Hilmert H, Prinz H, Heukeshoven J, Beyreuther K (1985) The protein component of scrapie-associated fibrils is a glycosylated low molecular weight protein. EMBO J 4: 1495–1501
Oesch B, Westaway D, Wälchli M, McKinley MP, Kent SBH, Aebersold R, Barry RA, Tempst P, Teplow DB, Hood LE, Prusiner SB, Weissmann C (1985) A cellular gene encodes scrapie PrP 27–30 protein. Cell 40: 735–746
Prusiner SB (1982) Novel proteinaceous infectious particles cause scrapie. Science 216: 136–144
Prusiner SB (1985) Prions: novel infectious pathogens. Adv Virus Res 29: 1–56
Prusiner SB, McKinley MP, Bowman KA, Bolton DC, Bendheim PE, Groth DF, Glenner GG (1983) Scrapie prions aggregate to form amyloid-like birefringent rods. Cell 35: 349–358
Virchow R (1854a) Über eine im Gehirn und Rückenmark des Menschen aufgefundene Substanz mit der chemischen Reaktion der Cellulose. Virchows Arch [Pathol Anat] 6: 135–138
Virchow R (1854b) Zur Cellulose-Frage. Virchows Arch [Pathol Anat] 6: 416–426
Westaway D, Prusiner SB (1986) Conservation of the cellular gene encoding the scrapie prion protein PrP 27–30. Nucleic Acids Res 14: 2035–2044

# Mechanism of Recovery from Acute Virus Infection*

Fritz Lehmann-Grube

Heinrich-Pette-Institut für Experimentelle Virologie und Immunologie an der Universität Hamburg, Martinistrasse 52, 2000 Hamburg 20, Federal Republic of Germany

## The System

Virology as we know it today is essentially "molecular virology" and primarily directed at studying the structure of viruses and their interaction with cells. This was not always so. When this field began to evolve in its own right, medical and biological phenomena were predominantly analyzed, and little attention was paid to events occurring at levels below the whole organism. Probably this emphasis on the biology of viruses was an expression of the then prevailing interest of scientists, but also a reflection of the available methodological repertoire, because, no sooner had this expanded to allow performance of experiments in the test tube, than virology became what it is at present. Probably the key event in this regard was the publication in 1949 by Enders and his colleagues in which it was reported that the poliovirus multiplied in and caused cytopathic destruction of cultivated cells not belonging to the nervous system (Enders et al. 1949).

In the more than 35 years that have passed since, much has been learned about viruses and the effect they have on cells in vitro, but our knowledge of the events taking place in higher organisms has not progressed at the same pace. With the now available basic information, in conjunction with newly developed methods that allow the study of viruses in their natural environment more effectively than had previously been possible, it appears promising to take up "biological virology" again and try to answer such questions as how an animal is infected and how the infection is terminated, where viruses multiply and how they cause illness, and why it is that these agents sometimes persist.

Probably even less would be known about the relationship between viruses and higher organisms if it were not for the LCM virus-infected mouse, the study of

* Work done by the author and his colleagues was aided by research grants from the Deutsche Forschungsgemeinschaft.

*Abbreviations:* AFC, cells forming antibodies against LCM virus; B6 mouse, C57BL/6J mouse; CBA mouse, CBA/J mouse; CTL, cytotoxic T lymphocyte; DS500, dextran sulfate with molecular weight 500000; i.c., intracerebral; i.p., intraperitoneal; IU, mouse infectious unit; i.v., intravenous; LCM, lymphocytic choriomeningitis; MHC, major histocompatibility gene complex; MNP, mononuclear phagocyte.

The Heinrich-Pette-Institut is financially supported by the Freie und Hansestadt Hamburg und by the Bundesministerium für Jugend, Familie, Frauen und Gesundheit.

which has contributed much to our understanding of, for instance, immunopathologic phenomena in virus diseases, virus-specific immunologic tolerance, and virus persistence, and even topics of biological importance outside the field of virology such as the restriction of immunologic recognition by products of the MHC (Zinkernagel and Doherty 1979) or the activity of natural killer cells (Welsh 1981) were influenced by findings with this virus and its natural host, the gray house mouse (Lehmann-Grube 1984).

The events that characterize the infection of mice with LCM virus are as follows. If the first encounter occurs in utero or soon after birth, the animals remain healthy, although a carrier state develops in which large quantities of virus persist throughout life in all organs (Traub 1939). If, on the other hand, the adult animal is infected experimentally, the outcome depends on mouse strain, strain and dose of virus, and route of inoculation, but differs in every case from what happens in unborn or newborn mice. The rule is that s.c. or i.v. inoculation has few or no pathological sequels, although the virus may multiply to high titers in all major organs, whereas i.c. inoculation essentially always, and i.p. inoculation often, results in characteristic illnesses

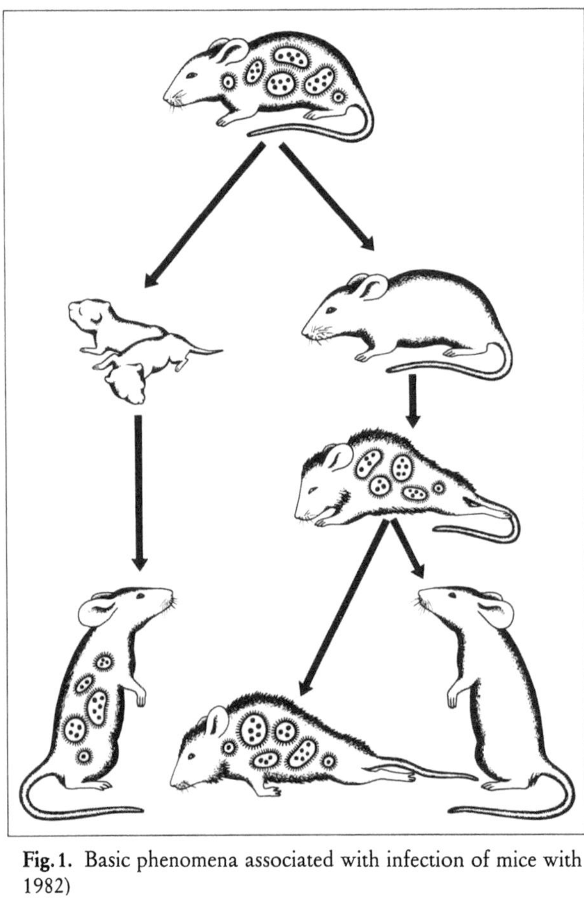

**Fig. 1.** Basic phenomena associated with infection of mice with LCM virus. (From Lehmann-Grube 1982)

that usually terminate in death. In surviving animals the virus is rapidly cleared. The basic phenomena characterizing the relationship between LCM virus and *Mus musculus* are depicted in Fig. 1. The explanation of this apparent paradox is as follows. The LCM virus is essentially harmless for the cells in which it multiplies and its lifelong persistence after prenatal or neonatal infection is possible because of virus-specific immunologic tolerance (Volkert and Hannover Larsen 1965). This tolerance is restricted to the T cell compartment (Lehmann-Grube et al. 1983); LCM virus-specific antibodies are often produced and are held responsible for a wasting disease from which carrier mice often suffer (Hotchin and Collins 1964; Oldstone 1975). Illness and death of acutely infected adult mice, on the other hand, result from an excessively vigorous interaction between virus or virus-induced new cellular antigens on one hand and the immune system on the other (Hotchin 1962). If the mouse survives, it controls the infection probably by the same mechanism that operates in other virus infections.

## The Model

Infection and (infectious) disease signify two entirely different phenomena, the former being invasion of the macroorganism or cell by an agent followed by its multiplication, the latter referring to the pathologic consequences thereof. The mechanisms governing an infection and its termination and the events leading to illness and restoration may overlap to some extent but are principally different. Thus, when studying the course of a virus infection in an experimental animal, illness and death may be conveniently registered parameters but say little about the state of the infection. Analysis of the latter requires localization and quantitation of the virus, which, of course, is more laborious than mere observation of the animal. Indeed, the ideal model would be one in which pathologic alterations are totally absent and hence not able to confound the infectious process. I am not aware of the existence of such a model, but the adult mouse infected i.v. with a moderate dose of strain WE LCM virus comes very close.

After i.v. inoculation of $10^3$ mouse IU the virus multiplies in all major organs. Maximum titers are reached within 4–6 days, after which the virus is rapidly eliminated. Throughout, the mice remain outwardly healthy, and histologic inspection reveals minor changes in most tissues, although inflammatory alterations may be prominent in the liver. We have begun our work by analyzing events in the effector phase, i.e., the time during which the virus is eliminated (as opposed to the induction phase), and have chosen the spleen because in this organ (1) virus replicates rapidly, attaining concentrations as high as almost $10^9$ IU/g tissue within 4–5 days, (2) elimination of the virus is as fast, the infectivity usually being below detectability on day 10, and (3) histologic alterations are minimal (Lehmann-Grube et al. 1985). We realize that all cells that may be assumed to play a role in virus elimination are assembled in this organ, which is not the case elsewhere in the body, but are convinced that this will have no bearing on the basic mechanism, although it may influence its time kinetics. We also study the events in other organs and so far have found no principal differences.

## Role of T Lymphocytes

It has long been known that T lymphocytes are a conditio sine qua non for terminating infection of the mouse with LCM virus; this has probably been demonstrated most convincingly by showing that mice genetically or surgically deprived of their thymuses are unable to get rid of the virus (Rowe et al. 1963; Christoffersen et al. 1976). T lymphocytes can be broadly subdivided into at least two subpopulations, one having the potential to lyse antigenically altered cells and to suppress immunologic activity, the second to help other elements of the immune system and to induce them to become active. Besides these functional differences, the subsets also vary with respect to cell surface properties, which makes it possible to separate them experimentally (Walker et al. 1984).

During infection, T lymphocytes appear in the spleen that have the ability to lyse LCM virus-infected cells in vitro. Their activity has been shown to be restricted by class I products of the MHC (Zinkernagel and Doherty 1979), and according to results of negative selection experiments these cells are Lyt-$2^+$, L3T$4^-$ (Assmann-Wischer et al. 1985/86), which is characteristic of CTL. The kinetics of appearance and disappearance of CTL in the murine spleen correlate well with virus clearance (Marker and Volkert 1973; Lehmann-Grube et al. 1985). Furthermore, adoptive immunization experiments, in which infusion of LCM-immune spleen cells into previously infected mice reduced virus replication in the recipients, revealed that the effector cells were class I MHC-restricted, Lyt-$2^+$, L3T$4^-$ T lymphocytes (Zinkernagel and Welsh 1976; Assmann-Wischer et al. 1985/86), and spleen virus was also reduced in mice adoptively immunized with cloned CTL having the same properties (Byrne and Oldstone 1984). Thus, ample evidence exists for the notion that CTL are required for the removal of LCM virus from the murine spleen. According to one widely held opinion, CTL are central for the control of a virus infection by doing in situ exactly what they are capable of doing in vitro, namely lyse virus-infected cells, thereby preventing further virus assembly; infectious virus already released is assumed to be removed by antibodies and MNP. It seems to me that this is a rather crude mechanism, and it would be surprising if Nature had not found a more elegant solution to the problem of protecting higher organisms against viruses. This, of course, is a very personal view, but perhaps more convincing is a series of experiments the results of which are difficult to reconcile with the assumption that cytolysis is the mechanism that terminates a virus infection in vivo (Lehmann-Grube et al. 1985).

Four mouse strains were employed that varied as to the activity of LCM virus-specific spleen CTL arising during infection: DBA/1 and B6 are high-responder mice, and CBA and AKR are low-responder mice. Although the differences were considerable (Fig. 2), the virus was eliminated with closely similar speeds (Fig. 3), indicating that the CTL in these mice's organs were not rate limiting. This result led to the question of how many cells are actually needed to clear the virus. The answer was sought by adoptive immunization experiments, the results of which, depicted in Fig. 4, revealed marked differences between high-responder and low-responder mice with respect to the numbers of cells that were required to block virus replication in the recipients; more important, unexpectedly few were found to be thus effective. For instance, in the high-responder B6 mouse, approximately $10^4$ day-7-immune

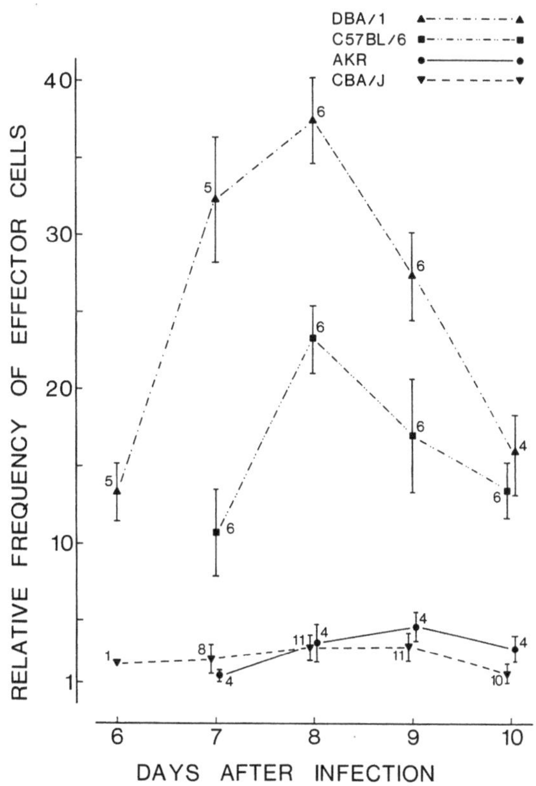


**Fig. 2.** CTL responses of mice of four strains determined with the chromium-release assay of Miller and Dunkley (1974), adapted to enumerate virus-specific CTL by Moskophidis and Lehmann-Grube (1983). Graded numbers of spleen cells were taken at intervals after i.v. infection with $10^3$ IU, and constant numbers of LCM virus-infected and $^{51}Cr$-labeled syngeneic fibroblastic target cells were incubated for 2.5 h at 37 °C and subsequently for 1 h at 45 °C, the shift of temperature promoting the release of label from previously damaged target cells, yet preventing new cytolytic encounters. The specific fractional lysis *(p)* was then calculated for each spleen cell-target cell combination, and the estimates of *p* for different numbers of sensitized spleen cells *(N)* were plotted against *N* on a log-log scale. For each value falling on a straight line with slope *1*, the relative frequency of effector cells *(α)* was calculated according to the formula $\alpha = \ln(1-p)/Nt$, where *t* is the time of incubation in hours at 37 °C. From the *α* values thus determined, the means were calculated, and these were the data from which means and standard errors were derived. The figures along the curves denote the numbers of independent determinations on which calculations of both these values were based. To simplify plotting, all *α* values were multiplied by $10^8$. (From Lehmann-Grube et al. 1985)

(predominantly) T lymphocytes reduced virus replication in the spleens within 40 h by more than 50%. We also determined the distribution of the transfused cells in the organs of virus-infected mice and found that 20% homed to the spleen. Thus, roughly $2 \times 10^3$ T lymphocytes could have been locally active.

To answer the obvious question of how many of these were LCM virus-specific CTL, the method of limiting dilution was employed (Assmann-Wischer et al. 1985/86; Moskophidis et al. 1987a). B6 mice were infected by i.v. inoculation of

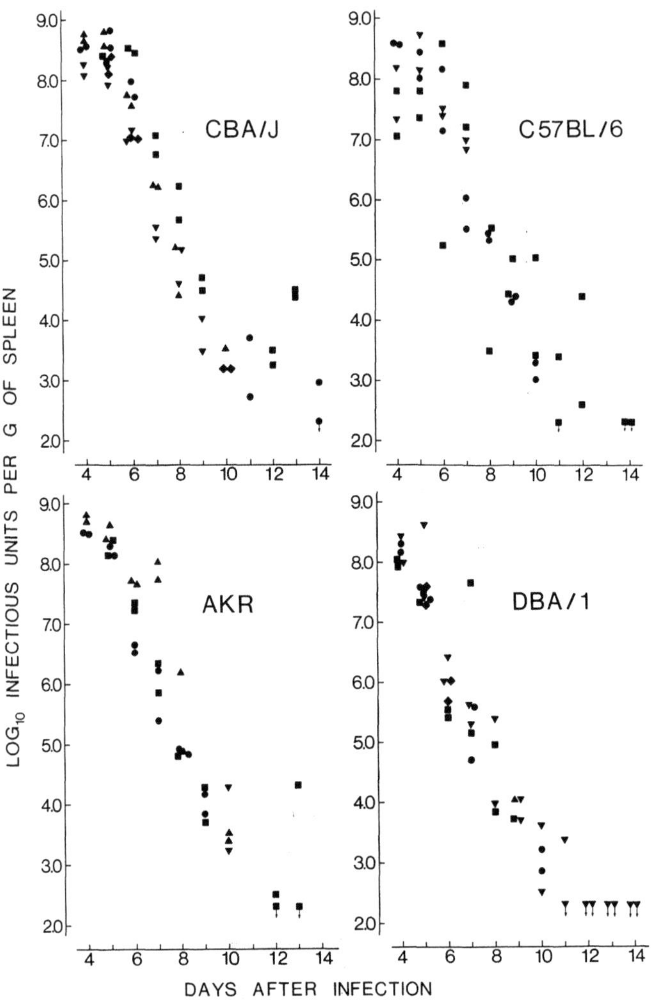


**Fig.3.** Rates of elimination of LCM virus from spleens of mice after i.v. infection with $10^3$ IU. Each symbol stands for the virus concentration of one organ, different shapes signifying different experiments. (From Lehmann-Grube et al. 1985)

$10^3$ IU, and at intervals defined numbers of spleen cells were cultivated in multiple wells of microculture plates together with virus-infected syngeneic stimulator cells and T cell growth factor, and after 7 days individual cultures were tested for ability to cause release of $^{51}Cr$ from infected and uninfected target cells. The results of one such experiment are presented in Table 1. In cultures seeded with spleen cells from uninfected mice or from mice infected for 3 days, no cytolytic activity was ever observed. On day 5, a few cells with the ability to develop into cytolytic clones had appeared. These cells rose rapidly in number, and on day 8 values as high as around one in 10 were calculated. In low-responder CBA mice the maximum frequencies of CTL and their precursors remained ca. 20 times lower. A small proportion of micro-

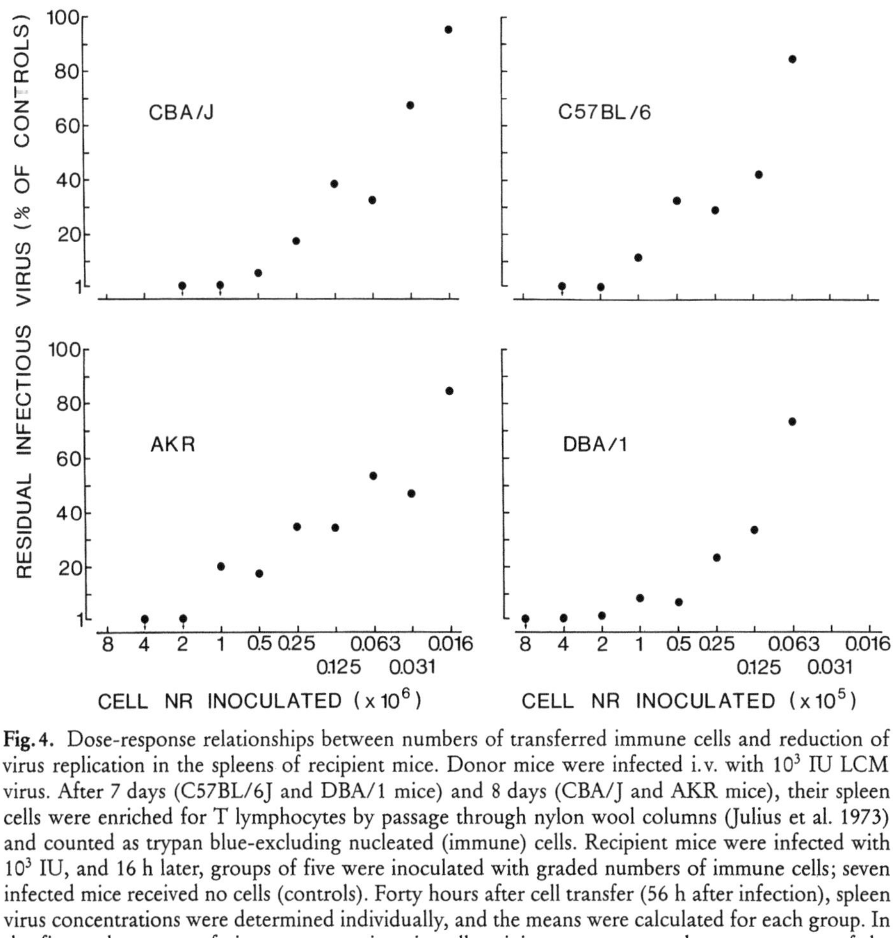


**Fig. 4.** Dose-response relationships between numbers of transferred immune cells and reduction of virus replication in the spleens of recipient mice. Donor mice were infected i.v. with $10^3$ IU LCM virus. After 7 days (C57BL/6J and DBA/1 mice) and 8 days (CBA/J and AKR mice), their spleen cells were enriched for T lymphocytes by passage through nylon wool columns (Julius et al. 1973) and counted as trypan blue-excluding nucleated (immune) cells. Recipient mice were infected with $10^3$ IU, and 16 h later, groups of five were inoculated with graded numbers of immune cells; seven infected mice received no cells (controls). Forty hours after cell transfer (56 h after infection), spleen virus concentrations were determined individually, and the means were calculated for each group. In the figure the means of virus concentrations in cell recipients are expressed as percentages of the means of virus concentrations in control mice. Note that the abscissa values for CBA/J and AKR mice *(left)* and C57BL/6J and DBA mice *(right)* differ by the factor 10. (From Lehmann-Grube et al. 1985)

cultures proved cytolytic also for noninfected syngeneic targets, but the dose-response relationships with numbers of seeded cells were irregular, and, furthermore, always considerably lower counts were set free from uninfected than from infected targets. Lysis of the majority of clones was restricted by products of the MHC; only a few lysed uninfected or infected allogeneic target cells. Positive and negative selection experiments disclosed that the responding cells were predominantly T lymphocytes expressing Lyt-2 but not L3T4 surface markers. Cells of a proportion of the cultures also lysed uninfected YAC cells. Presumably, this natural killer-like activity was acquired by T lymphocytes during their prolonged cultivation in vitro. We conclude that most spleen cells that during acute infection with LCM virus attained the ability to develop in vitro into cytolytic clones were derived from MHC-restricted, Lyt-$2^+$, L3T$4^-$ LCM virus-specific CTL and their activated precursors.

**Table 1.** Numbers of LCM virus-specific CTL and CTL precursors in spleens of acutely infected B6 mice

| Days after infection | CTL (P)[a] |
|---|---|
| 0 | <1/10$^5$ |
| 3 | <1/10$^5$ |
| 5 | 1/7590 |
| 6 | 1/80 |
| 7 | 1/48 |
| 8 | 1/10 |
| 9 | >1/25[b] |
| 10 | 1/62 |
| 14 | 1/50 |

From Assmann-Wischer et al. (1986)
[a] Frequency of cells that gave rise to clones lysing LCM virus-infected syngeneic targets
[b] At 78 cells per cup all cultures were positive

As to the numbers of cells that are required to diminish virus replication in the recipient's spleen, we had already come to the conclusion that in the case of the B6 mouse, 2000 spleen cells caused a reduction by at least 50%. Knowing now that 10% of these were LCM virus-specific CTL, we calculate that this effect must have been achieved by 200 such cells, which is a low figure, even if one concedes that in the 40 h between transfer and determination of virus titer some expansion of the CTL had probably occurred.

Doubts that these cells should have been capable of locating and destroying a large number of virus-infected cells in the spleen increase further if the time is considered within which the effect becomes apparent. By transfer of only a few more cells, the reduction of virus replication could be made evident much faster. The results of one experiment of this kind are detailed in Fig. 5. They show that $5 \times 10^5$ immune T lymphocytes diminished infectious virus in the recipients' spleens by more than 50% within 8 h, and by applying the same calculation as above, this must have been achieved by 10000 CTL. The time interval could be further shortened to 6 h by further increasing the number of infused cells (Table 2), and under these conditions it became obvious that the quantity of virus in the recipients' organs was reduced not only relative to the controls but also in absolute terms, suggesting that the immune cells had exerted some virucidal activity.

Any attempt to interpret these findings meets with difficulties. In all experiments 1000 IU were inoculated and the intervals between infection and transfer of immune cells were 16 h. Assuming that 10% of the virus reached the spleen (probably less) and that its replicative cycle in vivo is 14–16 h, as has been found to be the case in cultivated cells (Lehmann-Grube 1971), at the time the immune cells were inoculated, the spleen contained few potential target cells and these were likely to be focally distributed, an assumption that was confirmed immunohistologically by my colleague Jürgen Löhler (personal communication). Furthermore, virus-infected cells are not known to send signals; hence, if cytolysis in situ were the mechanism with which the infection is controlled, we would have to assume that a few CTL within a few hours had chance-encountered and destroyed 50% of virus-producing target

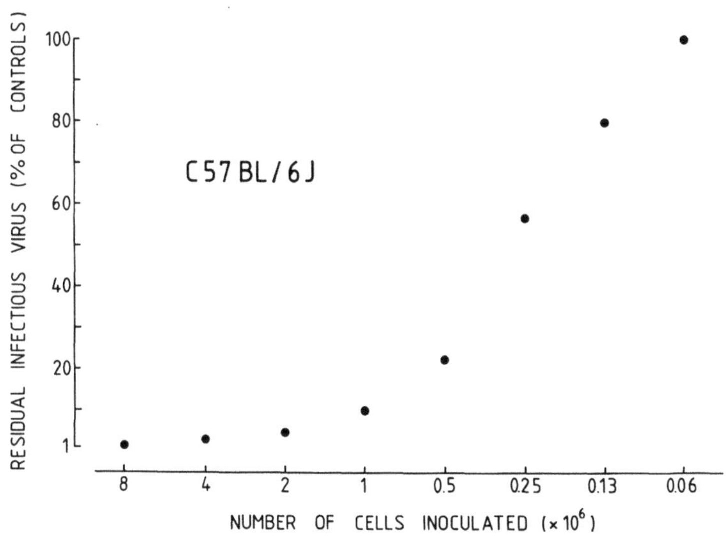


**Fig. 5.** Dose-response relationship between numbers of transferred immune T lymphocytes and reduction of virus concentration in the spleens of C57BL/6J recipient mice. Procedure as detailed for Fig. 4, except that infectious virus in recipient mice was quantitated 8 h after transfer of immune cells (24 h after infection). (From Assmann-Wischer et al. 1985/86)

**Table 2.** Effect after 6 h of transfer of variable numbers of day-8-immune spleen T lymphocytes on replication of LCM virus in spleens of B6 mice

| Time[a] (h) after | | Number of cells transferred | | | |
|---|---|---|---|---|---|
| Infection | Cell transfer | 0 (control) | $10^5$ | $10^6$ | $10^7$ |
| 16 | 0[b] | $2.7 \pm 0.7 \times 10^{3}$[c] | – | – | – |
| 22 | 6 | $4.7 \pm 1.6 \times 10^{4}$[c] | $4.3 \pm 1.2 \times 10^{4}$[c] | $\sim 4 \times 10^{3}$[c] | Trace |
| 46 | 30 | $6.6 \pm 1.5 \times 10^{6}$[d] | $2.0 \pm 0.5 \times 10^{6}$[d] | n. d. | n. d. |

From Lehmann-Grube et al. (1985) n. d., not done

[a] Sixteen hours after i.v. infection with $10^3$ IU, mice were adoptively immunized, and 6 h later (and 30 h later for control purposes) virus concentrations in spleens were determined

[b] No cells transferred

[c] Mean IU/g spleen ± standard error in seven mice

[d] Mean IU/g spleen ± standard error in three mice

cells assembled in a few foci in an organ consisting of approximately $2 \times 10^8$ cells. I find this difficult to accept and therefore favor an alternative explanation in which it is postulated that antiviral lymphokine(s) released by CTL diffuse into the surrounding tissue.

However, this interpretation, too, has its shortcomings. First, I know of no lymphokine that is capable to block virus replication in an already infected cell, although it seems to me that interferon $\gamma$ has not been adequately investigated in this respect. It is, of course, still possible that a very labile and therefore as yet unknown substance is involved. Secondly, as we have seen, the time interval required to make the effect apparent is brief, even if a diffusible substance rather than a cell is considered. Zinkernagel and Welsh (1976) have shown that the rules of MHC restriction also govern adoptive immunization of LCM virus-infected mice, and experiments of my own have disclosed that this applies even if the time interval between transfer and determination of virus replication is shortened to 8 h and the number of immune cells is greatly increased. Thus some cell-cell contact is required, unless the postulated factor is itself MHC restricted, for which there is no evidence. Thirdly, McIntyre and colleagues (1985) have performed experiments in which mice infected with LCM virus alone or the related Pichinde virus alone or both agents together were adoptively immunized with either LCM-immune or Pichinde-immune spleen cells. Only the virus that had been employed to immunize the donor mice was diminished in the recipients' spleens, which was interpreted to result from direct cytotoxicity or an extremely localized effect of soluble factors set free by the immune cells. Similar findings and a similar interpretation were reported by Lukacher et al. (1984), whose model is the lung of the influenza virus-infected mouse, but, there too, the last word has not been said. Under similar conditions, Schiltknecht and Ada (1985) recorded failure of adoptive immunization if, at the time of cell transfer, the recipients were under treatment with cyclosporin A, which to these authors was an indication that the infection was curtailed by some effect other than cytolysis, e.g., the action of lymphokines.

## Role of Mononuclear Phagocytes

Besides cytolysis by CTL, MNP, especially if activated, are widely assumed to represent effector cells of importance in the defence against virus infections (Mims 1964; Blanden et al. 1976; Mogensen 1979; Morahan et al. 1985). We have examined the general applicability of this supposition by determining the LCM virus' ability to replicate in the murine spleen, the MNP of which were either activated or blocked (Lehmann-Grube et al. 1986, 1987). This was assessed by the ability of *Listeria monocytogenes* to replicate during 24 h following i.v. inoculation and also by demonstrating histochemically the activity of enzymes that characterize these cells. In the LCM virus-infected spleen, especially during the time the infectious titer drops precipitously, the MNP were markedly activated (Fig. 6), which confirms and extends previous observations by Blanden and Mims (1973). Thus, it appeared quite possible that these cells played a role in the termination of the infection.

As already shown, virus clearance can be greatly accelerated by adoptive immunization, and we wondered whether activation of MNP would be essential in this process. The experimental answer was clearly negative. Rapid elimination of the virus from the spleen due to adoptive immunization was not accompanied by demonstrable activation of this organ's MNP, even if large numbers of immune T lymphocytes had been infused.

The MNP system comprises a heterogeneous group of cells with widely varying properties (Walker 1976; Adams and Hamilton 1984). For the present purpose they may be roughly subdivided into the monocytes and their precursors on one hand and the fully differentiated macrophages on the other. It is well established that for

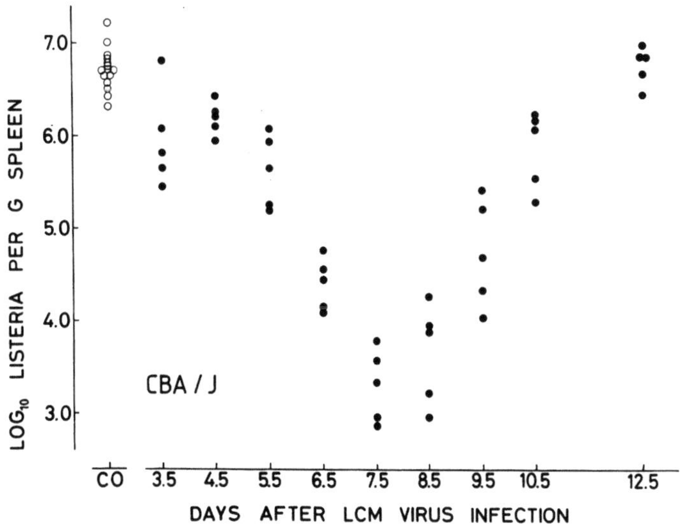


**Fig. 6.** Activation of MNP in the spleens of mice acutely infected with LCM virus. Mice were inoculated i.v. with $10^3$ IU of LCM virus and at intervals i.v. with $10^5$ colony-forming units of *Listeria monocytogenes* strain EGD. Twenty-four hours later the spleens were homogenized and viable *Listeria* enumerated on tryptic soy agar. (From Lehmann-Grube et al. 1986)

the immunologically mediated restraint of replication of *L. monocytogenes* in mouse organs at least two cell types are required, T lymphocytes and bone marrow-derived, rapidly multiplying, radio- and cyclophosphamide-sensitive elements of the MNP system, presumably monocyte precursors (Hahn and Kaufmann 1981). We wished to know whether the same holds true for our model. Mice were $\gamma$-irradiated and at intervals infected and adoptively immunized. Irrespective of irradiation dose (400–900 rad) and intervals till infection (2–10 days) the virus titers dropped as quickly in treated as in untreated mice. Pretreatment with cyclophosphamide led to the same negative results (Table 3), and we arrive at the conclusion that replication of *L. monocytogenes* and that of LCM virus in the murine spleen are controlled by fundamentally different mechanisms, at least as revealed by adoptive immunization.

Macrophages are highly resistant to X-irradiation (Geiger and Gallily 1974) and probably also to cyclophosphamide. Experiments were therefore performed in which these cells' function was likely to be also affected. In athymic nude mice the macrophages have been shown to be permanently activated (Rao et al. 1977; Mogensen and Andersen 1978), and in the spleens of these animals the virus replicated slightly but reproducibly less than in their euthymic counterparts. However, when the MNP (including macrophages) were activated by inoculating *Corynebacterium parvum* vaccine (Halpern et al. 1963; Milas and Scott 1978) prior to infection, the virus attained considerably higher concentrations than in untreated control mice (Table 4), but the elimination rate was not affected. Injection into mice of DS500 severely damages macrophages and related elements (Hahn and Bierther 1974). When mice thus treated were infected, the spleen virus titers were just as elevated as in mice the MNP of which had been activated with *C. parvum.* The subsequent elimination was

**Table 3.** Effect of cyclophosphamide on elimination of LCM virus from spleens of CBA mice adoptively immunized with syngeneic immune T lymphocytes

| No. of cells transferred[a] | No cyclophosphamide | Hours cyclophosphamide[b] | | | |
|---|---|---|---|---|---|
| | | −144 | −120 | −96 | −72 |
| 0 | $5.7 \pm 1.4 \times 10^7$ | $1.0 \pm 0.6 \times 10^7$ | $1.9 \pm 1.2 \times 10^7$ | $10.2 \pm 5.0 \times 10^7$ | $4.2 \pm 1.9 \times 10^7$ |
| $5 \times 10^5$ | $4.3 \pm 2.1 \times 10^5$ | $1.2 \pm 0.9 \times 10^5$ | $1.8 \pm 0.8 \times 10^5$ | $10.3 \pm 5.8 \times 10^5$ | $2.5 \pm 0.9 \times 10^5$ |

From Lehmann-Grube et al. (1986)
Data expressed in mean IU/g spleen ± standard error in five mice
[a] Donor mice were infected i.v. with $10^3$ IU. Eight days later their spleen cells were enriched for T lymphocytes by passage through nylon wool columns and counted as "living" on the basis of trypan blue exclusion. Recipient mice were treated with cyclophosphamide and at intervals thereafter were infected i.v. with $10^3$ IU. Sixteen hours after infection they were inoculated with day-8-immune T lymphocytes and 40 h later (56 h after infection) the virus concentrations in the spleens were determined
[b] Time in hours relative to cell transfer of i.p. inoculation of recipient mice with 200 mg/kg body weight cyclophosphamide

**Table 4.** Effect of activation of MNP with *Corynebacterium parvum* on the replication of LCM virus in the spleens of NMRI mice

| Treatment[a] | Days after infection | |
|---|---|---|
| | 4 | 5 |
| *C. parvum* | $51.9 \pm 9.5 \times 10^7$ | $65.2 \pm 20.1 \times 10^7$ |
| Control[b] | $9.4 \pm 4.2 \times 10^7$ | $5.4 \pm 2.1 \times 10^7$ |

Data expressed as mean IU/g spleen ± standard error in five mice
[a] Six days before i.v. infection with $10^3$ IU mice were inoculated i.v. with 1.4 mg *C. parvum* and at the indicated intervals virus concentrations in the spleen were determined
[b] Not treated with *C. parvum*

slightly delayed, but control experiments revealed that DS500 affected not only MNP but also CTL.

The essential finding of this study is the lack of a correlation between the functional state of the MNP system on one hand and replication and clearance of the virus on the other. At the time of spontaneous elimination the MNP were activated; this, however, was not the case when virus elimination followed adoptive immunization. In nu/nu mice with activated MNP the virus replicated slightly less than in nu/+ or +/+ controls, but virus production was increased in spleens of mice the MNP of which were either activated or damaged by treatment. The rate of elimination, on the other hand, was not noticeably influenced by these cells' state. Finally, depletion of monocytes in the recipients prior to adoptive immunization did not at all influence the accelerated virus clearance from the spleens. We think that these findings do not support any hypothesis in which MNP, whether activated or not, are assigned a role in the control of the LCM virus in the spleen of the mouse.

## Role of Antibodies

Besides CTL and MNP, B lymphocytes and their products are often assumed to play a part in controlling a virus infection. We have adapted a recently developed method for the enumeration of cells releasing antibodies of defined specificity (Sedgwick and Holt 1983) to localize and count AFC and have followed the development of AFC in the spleens of mice after i.v. infection with $10^3$ IU (Moskophidis and Lehmann-Grube 1984). The results (one representative experiment is detailed in Table 5) show that 5–10 days after infection, i.e., when virus is being eliminated, there is a rapid increase in numbers of both major classes of antiviral antibodies. Clearly, these findings would be compatible with the assumption that antibodies are involved in virus elimination. However, in mice the B cells of which had been depleted by treatment with antiserum directed against mouse IgM (Gordon 1979), the rate of virus elimination was only slightly retarded (F. Lehmann-Grube, unpublished material). More recently, we have determined the effect of treatment of mice with monoclonal antibodies against T lymphocytes and their subsets on virus elimination, CTL, and numbers of AFC (Moskophidis et al. 1987b). Whereas antibodies directed against Thy-1 (affecting all T lymphocytes) and Lyt-2 (affecting predominantly the cytolytic/suppressive subset of T lymphocytes) abolished both CTL activity and ability to clear the virus, they partially reduced (anti-Thy-1) or failed to reduce (anti-Lyt-2) numbers of AFC and, hence, production of antiviral antibodies. In contrast, treatment with antibody directed against L3T4 (affecting predominantly helper T lymphocytes) did not measurably influence the activity of either CTL or virus

**Table 5.** Numbers of spleen cells producing IgM and IgG anti-LCM virus antibodies during primary infection of CBA mice with $10^3$ IU LCM virus

| Days after infection | Numbers of antibody-producing cells | | | |
|---|---|---|---|---|
| | IgM | | IgG | |
| | Per $10^6$ cells[a] | Per spleen[b] | Per $10^6$ cells[a] | Per spleen[b] |
| 3 | 0 | – | 0 | – |
| 4 | 10 | 1185 | 0 | – |
| 5 | 21 | 2489 | 0 | – |
| 6 | 131 | 11888 | 125 | 11344 |
| 7 | 1033 | 122411 | 950 | 112575 |
| 8 | 2800 | 285600 | 3050 | 311100 |
| 9 | 583 | 117183 | 2450 | 492450 |
| 10 | 229 | 38472 | 1017 | 170856 |
| 11 | 114 | 27360 | 400 | 96000 |
| 12 | 119 | 29274 | 533 | 131118 |
| 18 | 10 | 1020 | 133 | 13566 |
| 23 | 15 | 2273 | 96 | 14544 |
| 38 | 14 | 1330 | 14 | 1330 |

From Moskophidis and Lehmann-Grube (1984)
[a] Counted as trypan blue-excluding ("variable") cells
[b] Counted as total numbers of cells using Türk's solution

clearance, yet abolished antiviral antibody production completely. Besides confirming previous conclusions concerning the central role of the Thy-1$^+$, Lyt-2$^+$, L3T4$^-$ CTL for virus clearance, these findings show that antibodies do not contribute measurably to the control of this primary infection.

## Concluding Remarks

Our aim being the elucidation of the mechanism by which higher organisms control viruses, we have chosen the LCM virus-infected mouse. In this report I have concentrated on the spleen, but further findings indicate that it may be permissible to extrapolate to other organs. I also think that our observations are of relevance not only with regard to this model; it appears unlikely that Nature should have invented different mechanisms to control different viruses. Be this as it may, at the present stage of our work I can state that of the components assumed to be part of the mechanism only CTL were found to be relevant. Indeed, all observations are compatible with the assumption that CTL alone are required to clear the LCM virus from the murine spleen. We do not yet know how they act, but it appears unlikely that cytolysis in situ is the mechanism. More probably, lymphokines block replication of the virus directly or through activation of other factor-producing cells.

## References

Adams DO, Hamilton TA (1984) The cell biology of macrophage activation. Annu Rev Immunol 2: 283–318

Assmann-Wischer U, Simon MM, Lehmann-Grube F (1985/86) Mechanism of recovery from acute virus infection. III. Subclass of T lymphocytes mediating clearance of lymphocytic choriomeningitis virus from the spleens of mice. Med Microbiol Immunol 174: 249–256

Assmann-Wischer U, Moskophidis D, Simon MM, Lehmann-Grube F (1986) Numbers of cytolytic T lymphocytes (CTL) and CTL precursor cells in spleens of mice acutely infected with lymphocytic choriomeningitis virus. Med Microbiol Immunol 175: 141–143

Blanden RV, Mims CA (1973) Macrophage activation in mice infected with ectromelia or lymphocytic choriomeningitis viruses. Aust J Exp Biol Med Sc 51: 393–398

Blanden RV, Hapel AJ, Doherty PC, Zinkernagel RM (1976) Lymphocyte-macrophage interactions and macrophage activation in the expression of antimicrobial immunity in vivo. In: Nelson DS (ed) Immunobiology of the macrophage. Academic, New York, pp 367–400

Byrne JA, Oldstone MBA (1984) Biology of cloned cytotoxic T lymphocytes specific for lymphocytic choriomeningitis virus: clearance of virus in vivo. J Virol 51: 682–686

Christoffersen PJ, Volkert M, Rygaard J (1976) Immunological unresponsiveness of nude mice to LCM virus infection. Acta Pathol Microbiol Scand [C] 84: 520–523

Enders JF, Weller TH, Robbins FC (1949) Cultivation of the Lansing strain of poliomyelitis virus in cultures of various human embryonic tissues. Science 109: 85–87

Geiger B, Gallily R (1974) Effects of X-irradiation on various functions of murine macrophages. Clin Exp Immunol 16: 643–655

Gordon J (1979) The B lymphocyte-deprived mouse as a tool in immunobiology. J Immunol Methods 25: 227–238

Hahn H, Bierther M (1974) Morphological changes induced by dextran sulfate 500 in mononuclear phagocytes of *Listeria*-infected mice. Infect Immun 10: 1110–1119

Hahn H, Kaufmann SHE (1981) The role of cell-mediated immunity in bacterial infections. Rev Infect Dis 3: 1221–1250
Halpern BN, Prévot AR, Biozzi G, Stiffel C, Mouton D, Morard JC, Bouthillier Y, Decreusefond C (1963) Stimulation de l'activité phagocytaire du système réticuloendothélial provoquée par *Corynebacterium parvum*. J Reticuloendothel Soc 1: 77–96
Hotchin J (1962) The biology of lymphocytic choriomeningitis infection: virus-induced immune disease. Cold Spring Harbor Symp Quant Biol 27: 479–499
Hotchin J, Collins DN (1964) Glomerulonephritis and late onset disease of mice following neonatal virus infection. Nature 203: 1357–1359
Julius MH, Simpson E, Herzenberg LA (1973) A rapid method for the isolation of functional thymus-derived murine lymphocytes. Eur J Immunol 3: 645–649
Lehmann-Grube F (1971) Lymphocytic choriomeningitis virus. Virology monographs 10. Springer, Vienna New York
Lehmann-Grube F (1982) Lymphocytic choriomeningitis virus. In: Foster HL, Small JD, Fox JG (eds) The mouse in biomedical research. Vol II. Diseases. Academic, New York, pp 231–266
Lehmann-Grube F (1984) Portraits of viruses: arenaviruses. Intervirology 22: 121–145
Lehmann-Grube F, Martínez Peralta L, Bruns M, Löhler J (1983) Persistent infection of mice with the lymphocytic choriomeningitis virus. Compr Virol 18: 43–103
Lehmann-Grube F, Assman U, Löliger C, Moskophidis D, Löhler J (1985) Mechanism of recovery from acute virus infection. I. Role of T lymphocytes in the clearance of lymphocytic choriomeningitis virus from spleens of mice. J Immunol 134: 608–615
Lehmann-Grube F, Assmann-Wischer U, Schwachenwald R, Krenz I, Krahnert T, Moskophidis D (1986) Questionable role of mononuclear phagocytes in the elimination of lymphocytic choriomeningits virus from spleens of acutely infected mice. Med Microbiol Immunol 175: 145–148
Lehmann-Grube F, Krenz I, Krahnert T, Schwachenwald R, Moskophidis D, Löhler J, Villeda Posada CJ (1987) Mechanism of recovery from acute virus infection. IV. Questionable role of mononuclear phagocytes in the clearance of lymphocytic choriomeningitis virus from spleens of mice. J Immunol 138: 2282–2289
Lukacher AE, Braciale VL, Braciale TJ (1984) In vivo effector function of influenza virus-specific cytotoxic T lymphocyte clones is highly specific. J Exp Med 160: 814–826
Marker O, Volkert M (1973) Studies on cell-mediated immunity to lymphocytic choriomeningitis virus in mice. J Exp Med 137: 1511–1525
McIntyre KW, Bukowski JF, Welsh RM (1985) Exquisite specificity of adoptive immunization in arenavirus-infected mice. Antiviral Res 5: 299–305
Milas L, Scott MT (1978) Antitumor activity of *Corynebacterium parvum*. Adv Cancer Res 26: 257–306
Miller RG, Dunkley M (1974) Quantitative analysis of the $^{51}Cr$ release cytotoxicity assay for cytotoxic lymphocytes. Cell Immunol 14: 284–302
Mims CA (1964) Aspects of the pathogenesis of virus diseases. Bacteriol Rev 28: 30–71
Mogensen SC (1979) Role of macrophages in natural resistance to virus infections. Microbiol Rev 43: 1–26
Mogensen SC, Andersen HK (1978) Role of activated macrophages in resistance of congenitally athymic nude mice to hepatitis induced by herpes simplex virus type 2. Infect Immun 19: 792–798
Morahan PS, Connor JR, Leary KR (1985) Viruses and the versatile macrophage. Br Med Bull 41: 15–21
Moskophidis D, Lehmann-Grube F (1983) The immune response of the mouse to lymphocytic choriomeningitis virus. III. Differences of numbers of cytotoxic T lymphocytes in spleens of mice of different strains. Cell Immunol 77: 279–289
Moskophidis D, Lehmann-Grube F (1984) The immune response of the mouse to lymphocytic choriomeningitis virus. IV. Enumeration of antibody-producing cells in spleens during acute and persistent infection. J Immunol 133: 3366–3370
Moskophidis D, Assmann-Wischer U, Simon MM, Lehmann-Grube F (1987a) The immune response of the mouse to lymphocytic choriomeningitis virus. V. High numbers of cytolytic T lymphocytes are generated in the spleen during acute infection. Europ J Immunol 17: 937–942
Moskophidis D, Cobbold SP, Waldmann H, Lehmann-Grube F (1987b) Mechanism of recovery from acute virus infection: treatment of lymphocytic choriomeningitis virus-infected mice with monoclonal antibodies reveals that Lyt-$2^+$ T lymphocytes mediate clearance of virus and regulate the antiviral antibody response. J Virol 61: 1867–1874

Oldstone MBA (1975) Virus neutralization and virus-induced immune complex disease. Progr Med Virol 19: 84–119

Rao GR, Rawls WE, Perey DYE, Tompkins WAF (1977) Macrophage activation in congenitally athymic mice raised under conventional or germ-free conditions. J Reticuloendothel Soc 21: 13–20

Rowe WP, Black PH, Levey RH (1963) Protective effect of neonatal thymectomy on mouse LCM infection. Proc Soc Exp Biol Med 114: 248–251

Schiltknecht E, Ada GL (1985) Influenza virus-specific T cells fail to reduce lung virus titres in cyclosporin-treated, infected mice. Scand J Immunol 22: 99–103

Sedgwick JD, Holt PG (1983) A solid-phase immunoenzymatic technique for the enumeration of specific antibody-secreting cells. J Immunol Methods 57: 301–309

Traub E (1939) Epidemiology of lymphocytic choriomeningitis in a mouse stock observed for four years. J Exp Med 69: 801–817

Volkert M, Hannover Larsen J (1965) Immunological tolerance to viruses. Prog Med Virol 7: 160–207

Walker WS (1976) Functional heterogeneity of macrophages. In: Nelson DS (ed) Immunobiology of the macrophage. Academic, New York, pp 91–110

Walker ID, Hogarth PM, Murray BJ, Lovering KE, Classon BJ, Chambers GW, McKenzie IFC (1984) Ly antigens associated with T cell recognition and effector function. Immunol Rev 82: 47–77

Welsh RM (1981) Natural cell-mediated immunity during viral infections. In: Haller O (ed) Natural resistance to tumors and viruses. Springer Berlin Heidelberg New York, pp 83–104 (Current topics in microbiology and immunology, vol 92)

Zinkernagel RM, Doherty PC (1979) MHC-restricted cytotoxic T cells: studies on the biological role of polymorphic major transplantation antigens determining T-cell restriction-specificity, function, and responsiveness. Adv Immunol 27: 51–177

Zinkernagel RM, Welsh RM (1976) H-2 compatibility requirement for virus-specific T cell-mediated effector functions in vivo. I. Specificity of T cells conferring antiviral protection against lymphocytic choriomeningitis virus is associated with H-2K and H-2D. J Immunol 117: 1495–1502

# Coronavirus Infection of Rats: A Model for Virus-Induced Demyelinating Disease Associated with Autoimmunity*

H. Wege, P. Massa, R. Dörries, R. Watanabe, and V. ter Meulen

Institut für Virologie und Immunbiologie der Universität Würzburg, Versbacher Strasse 7, 8700 Würzburg, Federal Republic of Germany

## Introduction

Several naturally occurring central nervous system (CNS) diseases of man and animals reveal lesions of primary demyelination in association with mononuclear cell infiltrations. Such diseases can develop after infections with viruses like measles, herpes or canine distemper. In some CNS diseases with unknown etiology, e.g., multiple sclerosis, a viral agent may be involved (Wisniewski et al. 1985). Circumstantial evidence suggests that the development of these CNS disorders could be mediated by immune reactions against host tissue (Wege et al. 1988a). The mechanisms by which a virus can trigger autoimmune reactions are not well understood. Therefore, experimental models allowing study of this aspect are of interest.

One of the most useful models to analyze mechanisms of demyelination is that of encephalomyelitides of rodents induced by infection with coronaviruses (Knobler et al. 1982; Wege et al. 1982, 1983a; Lavi et al. 1984). In this chapter we summarize our studies using coronavirus JHM in rats. The results clearly indicate that a persistent virus infection of glial cells can trigger an autoimmune reaction leading to damage of myelin.

## Virus and Host Factors Leading to Demyelination

Intracerebral infection of Lewis rats with JHM virus can lead to encephalomyelitis (Nagashima et al. 1978, 1979; Sörensen et al. 1980). Both the clinical course and the type of demyelinating lesions are influenced by host and viral factors (Table 1). Rats infected at the age of 8–12 days develop in most cases a rapidly progressing fatal paralytic disease, acute encephalomyelitis (AE), within a week p.i. Older rats come down either with AE or with a subacute demyelinating encephalomyelitis (SDE) which displays a slower onset of clinical signs (e.g., slight ataxic gait, paresis of hindlegs). Often the animals recover from disease or reveal chronic disabilities for several months. In some cases a relapse of clinical symptoms may occur after a long incubation period (Wege et al. 1984b). This SDE is based on a persistent virus infec-

* This work was supported by the Deutsche Forschungsgemeinschaft.

**Table 1.** Neurovirulence of JHM virus for Lewis rats

| Designation and type of virus | Age of rats at time of infection | |
|---|---|---|
| | 8–12 days | 4–5 weeks |
| JHM<br>Original strain passaged in vivo | AE | AE and SDE |
| JHM-Wt<br>Cell-adapted wild type | AE | AE |
| JHM-Ts 43<br>Temperature-sensitive mutant | AE and SDE | SDE |
| JHM-Pi<br>Released by persistently infected cells | Not virulent | Not virulent |

AE, acute encephalomyelitis; SDE, subacute demyelinating encephalomyelitis

tion of glial cells. Pronounced plaques of primary demyelination are distributed in areas of the white matter in association with infiltrations of lymphocytes and mononuclear cells. However, as shown in Table 1, the development of AE or SDE depends not only on the age of the animal but also on the biological properties of the virus.

Adaptation of JHM virus to tissue culture resulted in virus clones which are highly virulent. In contrast, a variant selected from persistently infected cells (JHM-Pi, Baybutt et al. 1984) is no longer neurovirulent. Temperature-sensitive mutants (e.g., JHM-Ts 43, Wege et al. 1983b) can induce high rates of SDE. These differences may depend on structural aspects of the E2 protein of this virus, as the following studies suggest.

## The Peplomer Protein as a Major Determinant of Neurovirulence

Coronavirus particles are composed of three major proteins. The genomic RNA is associated with the nucleocapsid protein (N). A lipid bilayer surrounds this entity and is protruded by a small glycoprotein (E1 or matrix). A large glycoprotein – termed E2 or peplomer – is anchored to the envelope and forms a corona of spikes. Similar to myxo- and paramyxoviruses, the capacity to fuse cells is activated by proteolytic cleavage of E2 into subunits (gp90A and B). Furthermore, epitopes on this protein elicit neutralizing antibodies and determine the host specifity of infection (Siddell et al. 1983; Sturman et al. 1985).

To analyze in detail the structural basis of the variable neurovirulence of JHM virus mutants, a panel of monoclonal antibodies (Mabs) against E2 protein was raised (Fleming et al. 1983; Wege et al. 1984a; Talbot et al. 1984). With these Mabs more than 10 epitopes were defined by a combination of biological and biochemical tests involving JHM mutants, related coronaviruses and variants which escape neutralization by Mabs (Wege et al. 1988b). As summarized in Table 2, four epitopes are recognized by antibodies which neutralize the virus (E2-Aa, E2-Ab, E2-Ba and E2-Bb). Two of these epitopes are associated with cell fusion (E2-Aa and E2-Ba). The JHM

**Table 2.** Reactivity of anti-E2 Mabs with JHM virus neurovirulence variants (neutralization and inhibition of cell fusion)

| Designation of epitope | JHM-Wt | | JHM-Ts43 | | JHM-Pi | |
|---|---|---|---|---|---|---|
| | Neutr. | Fus. inh. | Neutr. | Fus. inh. | Neutr. | Fus. inh. |
| E2-Aa | ++++ | ++++ | ++++ | ++++ | – | ++++ |
| E2-Ab | +++ | – | ++ | – | ++++ | – |
| E2-Ba | ++ | ++ | ++++ | ++ | – | – |
| E2-Bb | ++ | – | + | – | +++ | – |
| E2-C to E2-F | – | – | – | – | – | – |

Neutr., neutralisation assay; Fus. inh., inhibition of cell fusion

variants with different neurotropic properties can be clearly discriminated at the level of these E2 epitopes.

The domains of E2 responsible for binding of neutralizing antibodies are strongly overlapping and largely formed by the conformation of the protein. With the avirulent virus JHM-Pi the sites for neutralization can be discriminated from the site for cell fusion. This antigenic variability between variants is based on distinct structural changes at the level of the E2 protein subunits. Furthermore, the pathogenicity of JHM virus can be changed by selection with Mabs binding to defined epitopes (Fleming et al. 1986).

The different biological behavior of these variants could be due to changes in target cell specificity or cytopathogenicity for brain cells. It has been observed that passage of JHM virus in rat brain results in selection of variants with structural changes in the E2 protein (Taguchi et al. 1985). To obtain more insight into these mechanisms, we compared the tropism of these JHM viruses for neural cells.

## Selective Tropism of JHM Variants for Astrocytes and Microglia

Mixed glial cell cultures were established from brains of newborn Lewis rats. The cells mature and differentiate during culture for several weeks similar to the development at the level of the organism. As demonstrated in Fig. 1, the major components are type I astrocytes, microglia cells and oligodendrocytes. These cultures can be infected by JHM-Wt virus (Massa et al. 1986b). Within 1 day p.i., astrocytes and microglia cells display viral antigen, whereas oligodendrocytes are not infected. Within a week, large syncytia of infected astrocytes are spread throughout the culture (Fig. 2). Oligodendrocytes and neurons, however, remain largely uninfected. Infected oligodendrocytes are detectable only in association with fused astrocytes and not in the early stages of infection. Neurons and oligodendrocytes could be additional important targets in acute disease or may harbor the virus without productive infection (Knobler et al. 1981; Beushausen and Dales 1985).

Infection of such mixed cultures with JHM-Ts43 and JHM-Pi has a very different outcome. Again, astrocytes and microglia are the major target for virus replication. However, JHM-Ts43 induces significantly smaller syncytia than JHM-Wt, and a persistent infection with continuous release of infectious virus develops. Infection

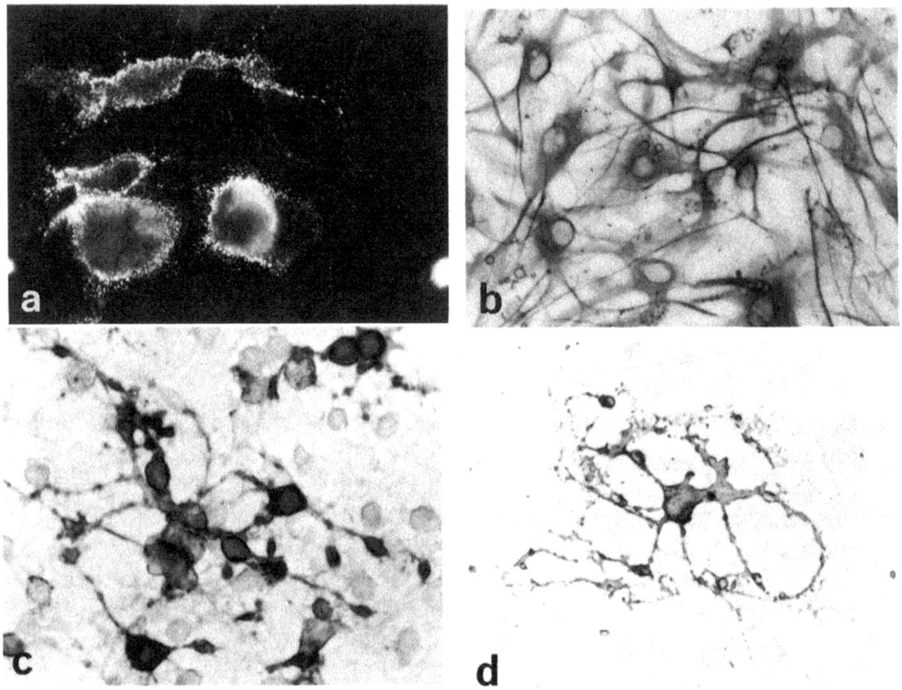

**Fig. 1 a–d.** Immunocytochemical staining of characteristic cell types in Lewis primary brain cultures (Massa et al. 1986 b). **a** Brain macrophages positive for Fc receptors; **b** type I and II astrocytes stained for glial fibrillary acidic protein (GFAP); **c** mature oligodendrocytes stained for myelin basic protein; **d** early oligodendrocyte stained for galactocerebroside

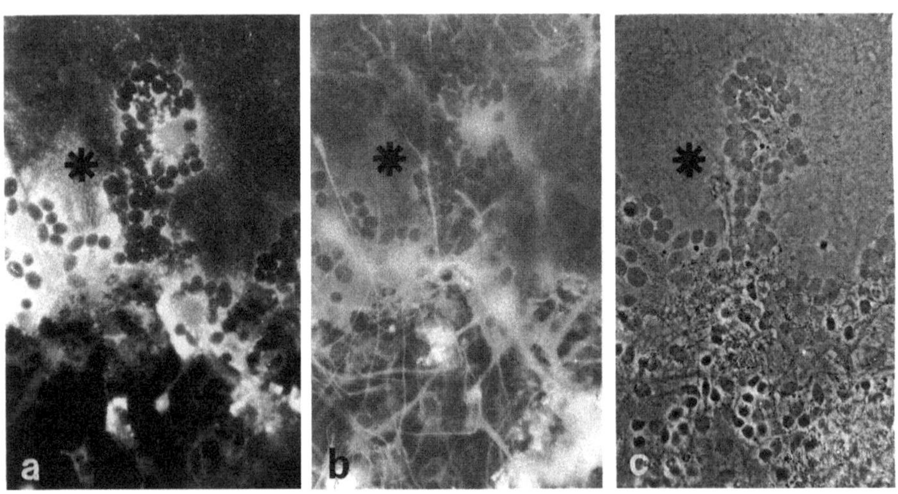

**Fig. 2 a–c.** Syncytia *(asterisk)* induced by JHM infection of astrocytes (double immune fluorescence). **a** Stained for JHM virus with a Mab against nucleocapsid protein; **b** the same field stained for GFAP; **c** same field, phase contrast

with the avirulent JHM-Pi leads either to an abortive infection of astrocytes or – at high input multiplicity – to a persistent infection without any visible cytopathology (Massa et al.1988). In contrast, in permanent cell cultures the same virus reveals similar growth kinetics to JHM-Wt and induces the formation of syncytia. Therefore, the cytopathogenicity specific for astrocytes could be an important determinant for the highly variable neurovirulence of coronaviruses.

### Host Genes Influence the Development of Disease

Several studies with inbred mice indicated that resistance to JHM virus infection is determined by host genes. Similar results were also obtained with inbred rat strains (Sörensen et al. 1982). We compared a collection of 15 inbred rat strains with regard to development of SDE. Brown Norwegian (BN) rats do not develop clinical signs of SDE, but JHM virus can establish a persistent infection in these rats (Watanabe et al. 1987). Neuropathological changes consist of nodular demyelinating lesions (Table 3). The lesions are confined to the periventricular area of the brain and characterized by extensive infiltrations of plasma cells and macrophages.

Congenic rats which contain the RT1 locus from BN rats on a Lewis background are still highly susceptible for disease. Since the RT1 genes of rats are the equivalent to the major histocompatibility complex region of other species (H2 in mice or HLA in humans), genes outside this region must determine susceptibility to disease. The different outcome of infection and the difference observed between CNS lesions in Lewis and BN rats suggest that host genes associated with immune regulation may influence the type of demyelination. The following results indicate that autoimmune reactions are important for development of SDE with pronounced clinical symptoms.

## Autoimmunity and the Pathogenesis of Demyelinating Disease

### Sensitization of Lymphocytes Against Myelin Basic Protein

The demyelinating lesions and perivascular infiltrations observed in Lewis rats with SDE resemble to some extent the changes seen in experimental allergic encephalomyelitis (EAE). EAE is induced by immunization with mixtures of adjuvants and myelin or myelin basic protein (MBP). The disease is largely promoted by a T cell-mediated delayed-type hypersensitivity reaction against brain tissue associated with perivenular cuffs of lymphocytes. Under certain conditions a chronic or relapsing EAE can be induced which results in demyelination (Wisniewski et al. 1985). In Lewis rats suffering from SDE we observed, in addition to immune reactions against virus, a T cell response against MBP (Watanabe et al. 1983). Lymphocytes from such rats proliferate in vitro in the presence of MBP to the same extent as observed for EAE. Moreover, transfer of such lymphocytes to normal rats by the intravenous route indeed resulted in clinical and pathological changes of EAE. In contrast, no sensitization of lymphocytes against MBP is observed in infected BN rats (Watanabe et al. 1987). This strain is also relatively resistant to induction of EAE.

**Table 3.** Course of coronavirus JHM infection in Lewis and BN rats

| Clinical signs | Lewis rats | BN rats |
|---|---|---|
| | ++ | – |
| **Histological changes** | | |
| Demyelination | ++ | ++ |
| Chronic inflammation | – | ++ |
| Spinal cord lesions | ++ | – |
| Perivascular infiltrations | ++ | +/– |
| Plasma cell infiltrations | +/– | ++ |

Age of rats at time of infection with JHM-Wt virus: 25–30 days

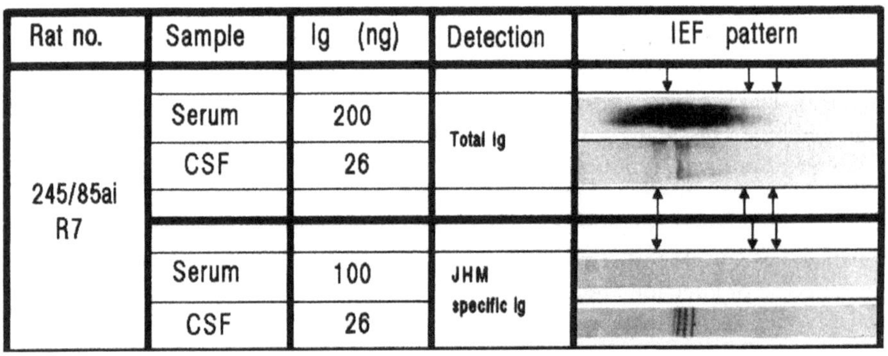


**Fig. 3.** Analysis of the Ig response by isoelectric focusing *(IEF)* and affinity-mediated immunoblots (Dörries et al. 1986a). Serum and CSF from a Lewis rat infected with JHM-Ts43 (dissected 4 months p.i.). The amount of Ig focused per track is indicated in the column "Ig (ng)." *Arrows* indicate the established pH gradient from left to right: pH 6.6, 8.2, 8.5. A JHM virus-specific Ig response of restricted heterogeneity is noticeable in the CSF specimen. These antibodies have been synthesized intrathecally, since the paired serum specimen does not reveal any virus-specific band, although five times as much Ig was focused per track

## Intrathecal Humoral Immune Response in Infected Lewis and BN Rats

The outcome of infection in Lewis and BN rats is also reflected by differences of humoral immune response in the CNS (Dörries et al. 1986a, b). The state of the blood-brain barrier and the immunoglobulin (Ig) synthesis was determined for paired serum and cerebral spinal fluid (CSF) samples. It could be shown that BN rats reveal a much stronger humoral immune response to JHM virus than Lewis rats. Furthermore, the specificity of Ig against virus and its clonal distribution was analyzed by isoelectric focusing in agarose gels combined with an immunoblot technique. As illustrated in Fig. 3, antiviral antibodies within the CSF can be of oligoclonal nature. In both Lewis and BN rats, intrathecally produced Ig is often detectable. However, in some Lewis rats additional Ig bands of restricted heterogeneity directed against unknown antigens were found. The strong antiviral immune response in BN rats in the absence of autoimmune reactions may help to prevent spread of virus and the development of clinical disease. These data suggest that a humoral immune response against viral antigens and neuroantigens could be maintained with the CNS despite the presence of an intact blood-brain barrier.

### Induction of Class II (Ia) Antigens by Virus Particles on Astrocytes

The question remains of how a persistent virus infection can trigger activation of autoimmune T cells which are involved in demyelination. Activation of T lymphocyte-mediated immune responses are in general triggered by cells which present the antigen after processing in context with class II glycoproteins (Ia antigens). In the CNS it has been shown that astrocytes can perform these functions if gamma-interferon is available (Hirsch et al. 1983; Fontana et al. 1984; Fierz et al. 1985). This substance is released by activated lymphocytes. However, activation of T cells does not take place within the brain compartment under normal conditions, since the intact blood-brain barrier restricts the traffic of lymphocytes and macromolecules. Therefore, in the case of a viral infection, induction of Ia antigen on astrocytes has to occur first, independent of gamma-interferon and T cells, to allow recognition of the antigen by invading lymphocytes. That viruses can induce Ia antigen on astrocytes has recently been demonstrated by us in rat brain cultures infected with JHM virus (Massa et al.1986a). This phenomenon was independent of viral replication in astrocytes, since ultraviolet-inactivated virus was also effective (Fig. 4). A neutralizing monoclonal antibody against the viral E2 glycoprotein prevented the induction of Ia, whereas a non-neutralizing antibody had no effect. These experiments suggest that either specific binding of the virus to the surface of astrocytes or phagocytosis of virus particles initiates expression of Ia. The potential of a neurotropic virus to induce Ia on astrocytes could have important pathogenetic implications for induction of an immune response against viral and host antigens.

## Conclusions

The results of studies of coronavirus infections in rodents are of increasing general interest for our understanding of mechanisms leading to chronic diseases of the CNS. The following conclusions can be drawn from these investigations:

1. The peplomer protein E2 is a major determinant for neurovirulence of coronaviruses. Distinct structural and antigenic changes of this protein are associated with differences in pathogenicity and virulence.
2. The initial cell targets in the CNS are astrocytes and brain macrophages. Oligodendrocytes and neurons are relatively resistant. The selective tropism and cytopathogenicity (especially cell fusion) for astrocytes is a major determinant for neurovirulence.
3. Viral particles induce the expression of class II (Ia) antigens on astrocytes. This induction is probably associated with binding of E2 protein to the cell surface,

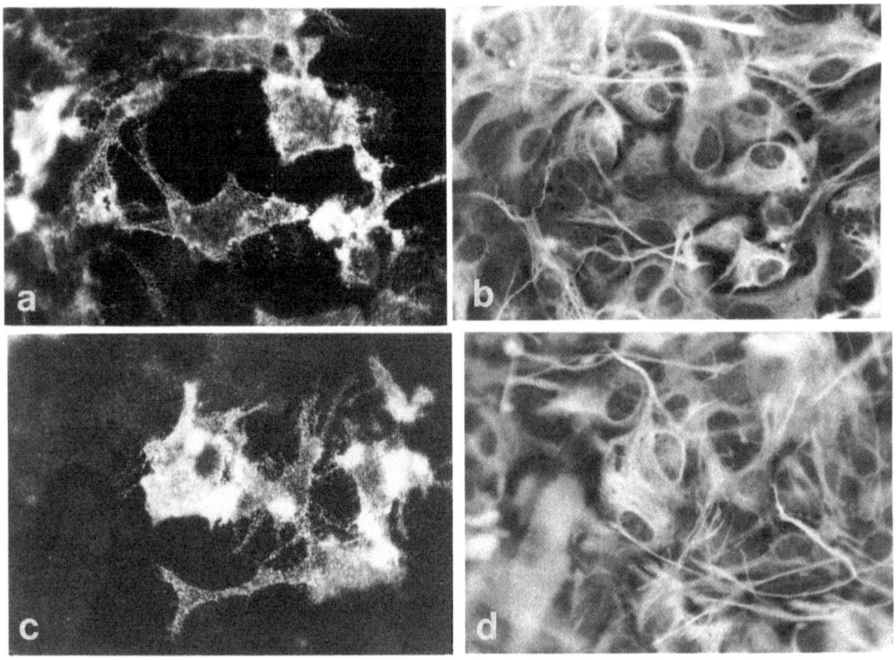


**Fig. 4 a–d.** Ia induction on a primary Lewis rat glial culture (10 days post plating): immunocytochemical staining by double immune fluorescence (Massa et al. 1986a). **a, b** Ia induction by pretreatment with gamma-interferon (18 h, 10 U $ml^{-1}$): **a** surface staining for Ia; **b** same field, astrocytes stained for GFAP. **c, d** Ia induction by pretreatment with JHM virus (UV-inactivated, equivalent dose $10^3$ PFU $m^{-1}$, 4 days): **c** surface staining for Ia; **d** same field, astrocytes stained for GFAP

which may activate a mechanism seen with bacterial endotoxins. Aberrant expression of class II antigens could be of central importance in the triggering of immune responses against host tissue (Londei et al. 1984).

4. The development of chronic demyelinating disease is promoted by autoimmune reactions against brain tissue linked to certain host genes.

Coronavirus JHM infections in rats represent a model for inflammatory demyelination of the CNS. The pathogenetic mechanisms can be studied from a molecular biological and immunological point of view, since the coronaviruses are well characterized and immunological markers for rat lymphocytes and brain cells exist. Experience gained from these studies will certainly influence our understanding of related human diseases.

*Acknowledgement.* We thank Helga Kriesinger for preparing the manuscript.

## References

Baybutt HN, Wege H, Carter MJ, ter Meulen V (1984) Adaptation of coronavirus JHM to persistent infection of murine Sac(-) cells. J Gen Virol 65: 915–924
Beushausen S, Dales S (1985) In vivo and in vitro models of demyelinating disease. XI. Tropism and differentiation regulate the infectious process of coronaviruses in primary explants of the rat CNS. Virology 141: 89–101
Dörries R, Wege H, Watanabe R, ter Meulen V (1986a) Murine coronavirus induced encephalomyelitides in rats: analysis of immunoglobulins and virus-specific antibodies in serum and cerebrospinal fluid. J Neuroimmunol 12: 131–142
Dörries R, Watanabe R, Wege H, ter Meulen V (1986b) Analysis of the intrathecal humoral-immune response in corona virus induced encephalomyelitis of rats. In: Lowenthal A, Rous J (eds) Cellular and humoral immunological components of cerebrospinal fluid in multiple sclerosis. Plenum, New York, pp 37–45
Fierz W, Endler B, Reske K, Wekerle H, Fontana A (1985) Astrocytes as antigen-presenting cells: I. Induction of Ia antigen expression on astrocytes by T cells via immune interferon and its effect on antigen presentation. J Immunol 134: 3785–3793
Fleming JO, Stohlman SA, Harmon RC, Lai MMC, Frelinger JA, Weiner LP (1983) Antigenic relationships of murine coronaviruses: analysis using monoclonal antibodies to JHM (MHV-4) virus.Virology 131: 296–307
Fleming JO, Trousdale MD, El-Zaatari FAK, Stohlman SA, Weiner LP (1986) Pathogenicity of antigenic variants of murine coronavirus JHM selected with monoclonal antibodies. J Virol 58: 869–875
Fontana A, Fierz W, Wekerle H (1984) Astrocytes present myelin basic protein to encephalitogenic T cell lines. Nature 307: 273–276
Hirsch MR, Wietzerbin J, Pierres M, Goridis C (1983) Expression of Ia antigens by cultured astrocytes treated with gamma-interferon. Neurosci Lett 41: 199–204
Knobler RL, Haspel MV, Oldstone MBA (1981) Mouse hepatitis virus type-4 (JHM strain) induced fatal nervous system disease. Part I: genetic control and the murine neuron as the susceptible site of disease. J Exp Med 133: 832–843
Knobler RL, Lampert PW, Oldstone MBA (1982) Virus persistence and recurring demyelination produced by a temperature-sensitive mutant of MHV-4. Nature 298: 279–280
Lavi E, Gilden DH, Wroblewska Z, Rorke LB, Weiss SR (1984) Experimental demyelination produced by the A59 strain of mouse hepatitis virus. Neurology 34: 597–603
Londei M, Lamb JR, Bottazzo GF, Feldmann M (1984) Epithelial cells expressing aberrant MHC class II determinants can present antigen to cloned human T cells. Nature 312: 639–641
Massa PT, Dörries R, ter Meulen V (1986a) Viral particles induce Ia antigen expression on astrocytes. Nature 320: 543–546
Massa PT, Wege H, ter Meulen V (1986b) Analysis of murine hepatitis virus (JHM strain) tropism toward Lewis rat glial cells in vitro: type I astrocytes and brain macrophages (microglia) are the primary glial cell targets. Lab Invest 55: 318–327
Massa PT, Wege H, ter Meulen V (1988) Growth pattern of various JHM coronavirus isolates in primary rat-glial cell cultures correlates with differing neurotropism in vivo. Virus Res 9: 133–144
Nagashima K, Wege H, Meyermann R, ter Meulen V (1978) Corona virus induced subacute demyelinating encephalomyelitis in rats. A morphological analysis. Acta Neuropathol 44: 63–70
Nagashima K, Wege H, Meyermann R, ter Meulen V (1979) Demyelinating encephalomyelitis induced by a long-term corona virus infection in rats. Acta Neuropathol 45: 205–213
Siddell SG, Wege H, ter Meulen V (1983) The biology of coronaviruses. J Gen Virol 64: 761–776
Sörensen O, Percy D, Dales S (1980) In vivo and in vitro models of demyelinating diseases. III. JHM virus infection of rats. Arch Neurol 37: 478–484
Sörensen O, Dugre R, Percy D, Dales S (1982) In vivo and in vitro models of demyelinating disease: endogenous factors influencing demyelinating disease caused by mouse hepatitis virus in rats and mice. Infect Immun 7: 1248–1260
Sturman LS, Ricard CS, Holmes KV (1985) Proteolytic cleavage of the E2 glycoprotein of murine coronavirus: activation of cell-fusing activity of virions by trypsin and separation of two different 90K cleavage fragments. J Virol 56: 904–911

Taguchi F, Siddell SG, Wege H, ter Meulen V (1985) Characterization of a variant virus selected in rat brain after infection by coronavirus mouse hepatitis virus JHM. J Virol 54: 429–435
Talbot PJ, Salmi AA, Knobler RL, Buchmeier MJ (1984) Topographical mapping of epitopes on the glycoproteins of murine hepatitis virus-4 (Strain JHM): correlation with biological activities. Virol 132: 250–260
Watanabe R, Wege H, ter Meulen V (1983) Adoptive transfer of EAE-like lesions by BMP stimulated lymphocytes from rats with coronavirus-induced demyelinating encephalomyelitis. Nature 305: 150–153
Watanabe R, Wege H, ter Meulen V (1987) Comparative analysis of coronavirus JHM induced demyelinating encephalomyelitis in Lewis and BN rats. Lab Invest 57: 375–384
Wege H, Siddell S, ter Meulen V (1982) The biology and pathogenesis of coronaviruses. Curr Top Microbiol Immunol 99: 164–199
Wege H, Watanabe R, ter Meulen V (1983 a) Murine coronavirus infection of the central nervous system in rats. In: Mims CA, Cuzner ML, Kelly RE (eds) Viruses and demyelinating diseases. Academic, New York, vol 7, pp 67–79
Wege H, Koga M, Watanabe R, Nagashima K, ter Meulen V (1983 b) Neurovirulence of murine coronavirus JHM temperature-sensitive mutants in rats. Infect Immun 39: 1316–1324
Wege H, Dörries R, Wege H (1984 a) Hybridoma antibodies to the murine coronavirus JHM: characterization of epitopes on the peplomer protein (E2). J Gen Virol 65: 1931–1942
Wege H, Watanabe R, ter Meulen V (1984 b) Relapsing subacute demyelinating encephalomyelitis in rats in the course of coronavirus JHM infection. J Neuroimmunol 6: 325–336
Wege H, Dörries R, Massa P, Watanabe R (1988 a) Autoimmunity and immunepathological aspects of virus diseases. In: Cohen J (ed) Perspectives on autoimmunity. CRC Press, Miami
Wege H, Winter J, Meyermann R (1988 b) The peplomer protein E2 of coronavirus JHM as a determinant of neurovirulence definition of critical epitopes by variant analysis. J Gen Virol 69: 87–98
Wisniewski HM, Schuller-Lewis GB, Mehta PD, Madrid RE, Lassmann H (1985) Pathogenetic aspects of multiple sclerosis and experimental models of inflammatory demyelination. Conc Immunopathol 2: 128–150
Wong GHW, Bartlett PF, Clark-Lewis I, McKimm-Breschkin JL, Schrader JW (1985) Interferon-gamma induces the expression of H-2 and Ia antigens on brain cells. J Neuroimmunol 7: 255–278

# Structural Proteins of the Murine Coronavirus MHV-JHM*

I. Schmidt, D. Ebner, M. A. Skinner, M. Pfleiderer, B. Schelle-Prinz, and S. G. Siddell

Institut für Virologie und Immunbiologie der Universität Würzburg, Versbacher Strasse 7, 8700 Würzburg, Federal Republic of Germany

## Introduction

The coronaviruses are a group of pleomorphic, enveloped viruses which replicate in the cytoplasm of vertebrate cells and are associated with diseases of economic importance (Siddell et al. 1983a). The murine hepatitis viruses (MHV) have been extensively used in the laboratory for the study of viral pathogenesis, in particular as a component in a model for demyelinating diseases of man (Knobler et al. 1982; Watanabe et al. 1983; Massa et al. 1986; Wege et al., this volume). The MHV genome is a single-stranded, positive-polarity RNA of approximately 18 kilobases. The genome encodes three major virion proteins, the nucleocapsid (N), the membrane (M or E1) and the surface (S or E2) proteins, as well as several non-structural proteins (Sturman and Holmes 1983; Siddell et al. 1983b).

The organization and expression of the MHV genome has been studied in some detail (for reviews see Holmes 1985; Siddell 1986). Briefly, in MHV-infected cells six subgenomic mRNAs, as well as genome-sized RNA, are produced. These mRNAs form a 3′ coterminal nested set and each has a common 5′ leader sequence of about 70 bases (Fig. 1) (Lai et al. 1981, 1984). The available evidence suggests that only the information contained within the "unique" sequences at the 5′ end of each mRNA (i.e., those absent from the next smallest RNA) is translated into protein (Siddell 1986). The in vitro translation of size-fractioned MHV mRNA has shown that the N protein is translated from mRNA 7, the M protein from mRNA 6 and the S protein from mRNA 3 (Siddell 1983). The nucleotide sequences of the genes encoding the MHV-JHM N, M and S proteins have recently been determined (Skinner and Siddell 1983; Pfleiderer et al. 1986; Schmidt et al. 1987). In this chapter we shall discuss the important features of the predicted amino acid sequences of these virion proteins.

* This work was supported by the Deutsche Forschungsgemeinschaft (SFB 165).

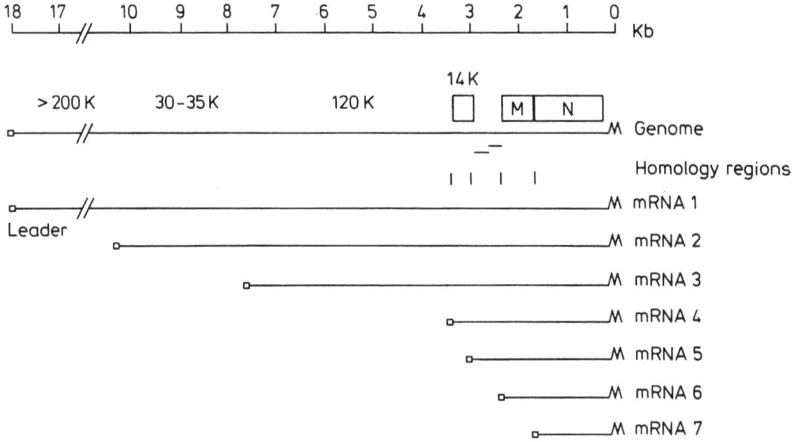


**Fig. 1.** The genomic organization of MHV. The relationship between the 3′ coterminal nested set of mRNAs and the viral genome is shown, together with the coding regions for the structural proteins, nucleocapsid *(N)*, membrane *(M)* and surface *(S)*, specified by mRNAs 7, 6 and 3 respectively

| | | | | | |
|---|---|---|---|---|---|
| 1 | MSFVPGQENA | GSRSSSGNRA | GNGILKKTTW | ADQTERGLNN | QNRGRKNQPK |
| 51 | QTATTQPNSG | SVVPHYSWFS | GITQFQKGKE | FQFAQGQGVP | IANGIPASQQ |
| 101 | KGYWYRHNRR | SFKTPDGQQK | QLLPRWYFYY | LGTGPYAGAE | YGDDIEGVVW |
| 151 | VASQQAETRT | SADIVERDPS | SHEAIPTRFA | PGTVLPQGFY | VEGSGRSAPA |
| 201 | SRSGSRPQSR | GPNNRARSSS | NQRQPASTVK | PDMAEEIAAL | VLAKLGKDAG |
| 251 | QPKQVTKQSA | KEVRQKILNK | PRQKRTPNKQ | CPVQQCFGKR | GPNQNFGGPE |
| 301 | MLKLGTSDPQ | FPILAELAPT | AGAFFFGSKL | ELVKKNSGGA | DGPTKDVYEL |
| 351 | QYSGAVRFDS | TLPGFETIMK | VLNENLNAYQ | NQDGGADVVS | PKPQRKRGTK |
| 401 | QKAQKDEVDN | VSVAKPKSSV | QRNVSRELTP | EDRSLLAQIL | DDGVVPDGLE |
| 451 | DDSNV | | | | |

**Fig. 2.** The predicted amino acid sequence of the MHV-JHM nucleocapsid protein

## Nucleocapsid Protein

The MHV-JHM nucleocapsid protein gene encodes a polypeptide of 455 amino acids with a molecular weight of 49700 (Fig. 2). From the charge (Fig. 3A) and hydropathicity (Fig. 3B) analyses of the predicted protein (Staden 1986) it is apparent that there are five major basic regions, two of which are located near the middle of the protein. Moreover, these basic regions correspond quite closely with hydrophilic domains. In contrast, the C-terminus of the protein is relatively acidic. It is also noteworthy that one amino acid, serine, appears to be clustered at positions Ser 194 to Ser 220; this region contains 24% of the total serine content (nine residues) within 6% of the coding sequence. This clustering also coincides with a basic region of the protein and may be significant in view of the fact that the MHV-JHM nucleocapsid protein is specifically phosphorylated at serine residues (Siddell et al. 1981).

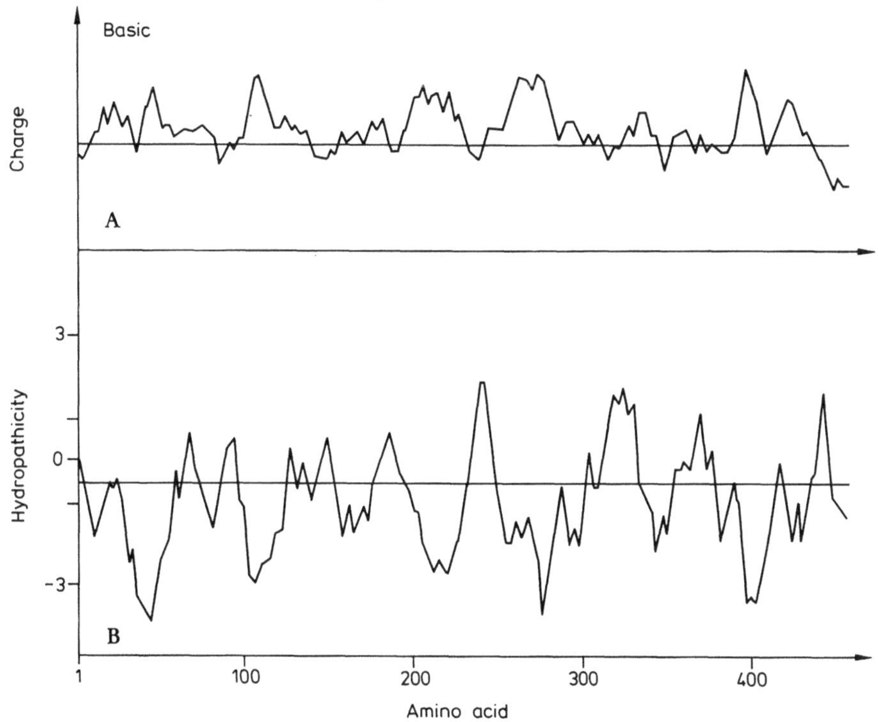


**Fig. 3 A, B.** **A** A charge analysis of the MHV-JHM nucleocapsid protein (Staden 1986). The analysis summates charge values over a window of nine amino acids. **B** A hydropathicity analysis of the MHV-JHM nucleocapsid protein according to Kyte and Doolittle (1982). The *vertical scale* is the average hydropathy for a frame of seven amino acids. The *midpoint line* represents the average hydropathy of the 20 amino acids. Hydrophobic sequences appear above this line

The nucleocapsid protein of MHV-JHM shows a very high homology (94%) with the nucleocapsid protein sequence of MHV-A59 (Armstrong et al. 1984a). More interestingly, as Boursnell et al. (1985) have pointed out, a striking degree of amino acid homology occurs between the MHV-JHM and IBV-Beaudette nucleocapsid protein sequences in a region of approximately 100 amino acids (amino acids 50–150 in the IBV protein, 85–185 in the MHV protein) in the N-terminal region of the polypeptides. As the RNA sequences encoding these proteins are essentially unrelated, this homology most probably represents an important functional domain. Also, there is a clustering of serine residues within the IBV protein at a position corresponding to the serine cluster in the MHV protein.

## Membrane Protein

The MHV-JHM membrane protein gene encodes a polypeptide of 228 residues with a molecular weight of 26 000 (Fig. 4). In comparison to the nucleotide sequence of the MHV-A59 M gene (Armstrong et al. 1984b) there are 21 nucleotide changes

```
  1 MSSTTQAPGP VYQWTADEAV QFLKEWNFSL GIILLFITII LQFGYTSRSM
 51 FIYVVKMIIL WLMWPLIIVL CMFNCVYALN NVYLGFSIVF TIVSVVMWIM
101 YFVNSIRLFI RTGSWWSFNP ETNNLMCIDM KGTVYVRPII EDYHTLTATI
151 IRGHFYMQGV KLGTGFSLSD LPAYVTVAKV SHLCTYKRAF LDKVDGVSGF
201 AVYVKSKVGN YRLPSNKPSG ADTVLLRI
```

**Fig. 4.** The predicted amino acid sequence of the MHV-JHM membrane protein

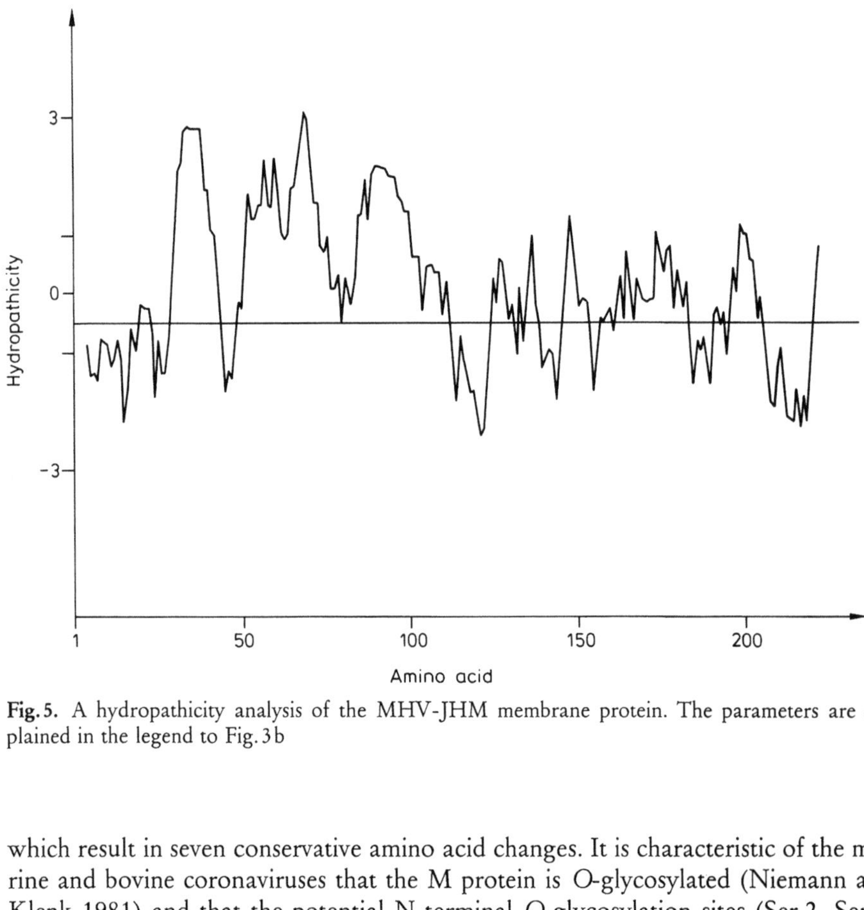


**Fig. 5.** A hydropathicity analysis of the MHV-JHM membrane protein. The parameters are explained in the legend to Fig. 3b

which result in seven conservative amino acid changes. It is characteristic of the murine and bovine coronaviruses that the M protein is O-glycosylated (Niemann and Klenk 1981) and that the potential N-terminal O-glycosylation sites (Ser 2, Ser 3, Thr 4, Thr 5) are conserved in the MHV-JHM sequence. Inspection of the hydropathicity analysis of this MHV-JHM protein (Fig. 5) reveals a characteristic N-terminal hydrophobic domain which appears to be comprised of three putative membrane spanning regions, as for MHV-A59, and the absence of a potential signal peptide (Rottier et al. 1984). Boursnell et al. (1984) compared the amino acid sequences of the MHV-A59 and IBV-Beaudette M proteins. A similar comparison between the MHV-JHM and IBV-Beaudette proteins, using the dot matrix comparison program DIAGON (Staden 1982), is shown in Fig. 6. This program, which looks not only for identical residues but also for residues with similar properties, re-

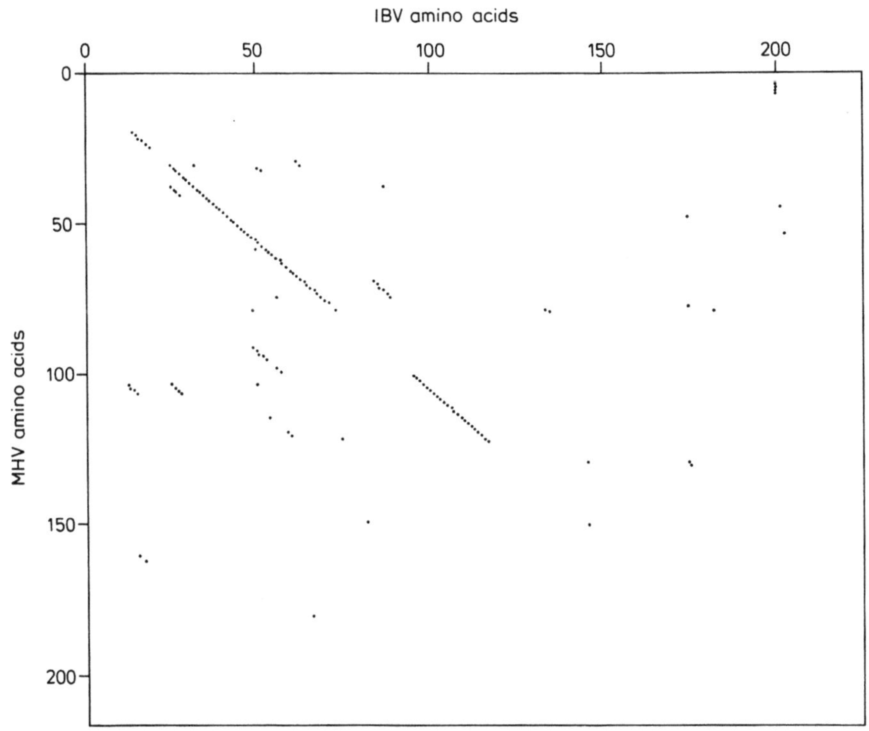


**Fig. 6.** Plot from DIAGON (Staden 1982) of the homologies between the MHV-JHM membrane protein sequence (from top to bottom) and the IBV-Beaudette membrane protein sequence (from left to right) (Boursnell et al. 1984). Points represent matches at the 1% significance level

veals a striking similarity in the N-terminal half of the proteins. To a large extent this reflects the hydrophobic nature of the putative membrane-spanning regions, but, as Boursnell et al. (1984) noted, the highest degree of similarity occurs in a hydrophilic region immediately "inside" the virion membrane.

## Surface Protein

The surface protein plays a central role in determining the cytopathic effects, virulence, tissue tropism and pathogenicity of coronavirus infections. The protein is responsible for the attachment and absorption of virions and is a major determinant for humoral and cell-mediated immune responses (Frana et al. 1985; Collins et al. 1982; Massa et al. 1986). The MHV-JHM S gene encodes a polypeptide of 1235 amino acids with a molecular weight of 136600 (Fig. 7). The apparent molecular weight of the polypeptide synthesized in infected cells is approximately 170000–180000 (Siddell 1982), which suggests that a considerable number of the 21 potential *N*-glycosylation sites (Asn-X-Thr/Ser; Neuberger et al. 1972) found within the primary sequence are used.

| | | | | | |
|---|---|---|---|---|---|
| 1 | MLFVFILLLP | SCLGYIGDFR | CIQTVNYNGN | NASAPSISTE | AVDVSKGRGT |
| 51 | YYVLDRVYLN | ATLLLTGYYP | VDGSNYRNLA | LTGTNTLSLT | WFKPPFLSEF |
| 101 | NDGIFAKVQN | LKTNTPTGAT | SYFPTIVIGS | LFGNTSYTVV | LEPYNNIIMA |
| 151 | SVCTYTICQL | PYTPCKPNTN | GNRVIGFWHT | DVKPPICLLK | RNFTFNVNAP |
| 201 | WLYFHFYQQG | GTFYAYYADK | PSATTFLFSV | YIGDILTQYF | VLPFICTPTA |
| 251 | GSTLAPLYWV | TPLLKRQYLF | NFNEKGVITS | AVDCASSYIS | EIKCKTQSLL |
| 301 | PSTGVYDLSG | YTVQPVGVVY | RRVPNLPDCK | IEEWLTAKSV | PSPLNWERRT |
| 351 | FQNCNFNLSS | LLRYVQAESL | SCNNIDASKV | YGMCFGSVSV | DKFAIPRSRQ |
| 401 | IDLQIGNSGF | LQTANYKIDT | AATSCQLYYS | LPKNNVTINN | YNPSSWNRRY |
| 451 | GFKVNDRCQI | FANILLNGIN | SGTTCSTDLQ | LPNTEVATGV | CVRYDLYGIT |
| 501 | GQGVFKEVKA | DYYNSWQALL | YDVNGNLNGF | RDLTTNKTYT | IRSCYSGRVS |
| 551 | AAYHKEAPEP | ALLYRNINCS | YVFTNNISRE | ENPLNYFDSY | LGCVVNADNR |
| 601 | TDEALPNCNL | RMGAGLCVDY | SKSRRARRSV | STGYRLTTFE | PYMPMLVNDS |
| 651 | VQSVGGLYEM | QIPTNFTIGH | HEEFIQIRAP | KVTIDCAAFV | CGDNAACRQQ |
| 701 | LVEYGSFCDN | VNAILNEVNN | LLDNMQLQVA | SALMQGVTIS | SRLPDGISGP |
| 751 | IDDINFSPLL | GCIGSTCAED | GNGPSAIRGR | SAIEDLLFDK | VKLSDVGFVE |
| 801 | AYNNCTGGQE | VRDLLCVQSF | NGIKVLPPVL | SESQISGYTA | GATAAAMFPP |
| 851 | WTAAAGVPFS | LNVQYRINGL | GVTMNVLSEN | QKMIASAFNN | ALGAIQEGFD |
| 901 | ATNSALGKIQ | SVVNANAEAL | NNLLNQLSNR | FGAISASLQE | ILTRLDAVEA |
| 951 | KAQIDRLING | RLTALNAYIS | KQLSDSTLIK | FSAAQAIEKV | NECVKSQTTR |
| 1001 | INFCGNGNHI | LSLVQNAPYG | LCFIHFSYVP | TSFKTANVSP | GLCISGDRGL |
| 1051 | APKAGYFVQD | NGEWKFTGSN | YYYPEPITDK | NSVAMISCAV | NYTKAPEVFL |
| 1101 | NNSIPNLPDF | KEELDKWFKN | QTSIAPDLSL | DFEKLNVTFL | DLTYEMNRIQ |
| 1151 | DAIKKLNESY | INLKEVGTYE | MYVKWPWYVW | LLIGLAGVAV | CVLLFFICCC |
| 1201 | TGCGSCCFRK | CGSCCDEYGG | HQDSIVIHNI | SAHED | |

**Fig. 7.** The predicted amino acid sequence of the MHV-JHM surface protein. The putative signal peptide, the transmembrane anchoring domain and the basically charged cleavage site are underlined

The predicted amino acid sequence of the MHV-JHM S propolypeptide is characteristic of group 1 membrane proteins (Garoff 1985). Analysis of the hydropathicity of the polypeptide reveals two regions of striking hydrophobicity. At the N-terminus there is a putative hydrophobic signal sequence which presumably is involved in insertion of the protein across the endoplasmic reticulum membrane during synthesis. At the C-terminus, there is a second hydrophobic domain which has the characteristics of a transmembrane anchoring region. This region is delineated from a hydrophilic "cytoplasmic" domain by two positively charged residues. On either side of this "charge cluster" there is a high proportion of cysteine residues, which may play a role in the acylation of this protein (Schmidt 1982; Rose et al. 1984).

During virus maturation the S protein is cleaved by a host cell enzyme to yield two approximately equally sized polypeptides $S_1$ and $S_2$ (Sturman et al. 1985). In the middle of the MHV-JHM sequence is a basically charged region which shows striking similarities to the known cleavage sites of glycoproteins of several other enveloped RNA viruses (White et al. 1983). In the case of MHV, it is not yet known whether the cleavage event is necessary for infectivity, but differences in the host cell proteolytic activity may well be important in determining permissivity.

The cleavage of the MHV S protein is also required to activate its fusogenic properties (Sturman et al. 1985). Examination of the MHV-JHM S protein sequence suggests that cleavage does not result in a hydrophobic domain at the N-terminus of $S_2$. This contrasts with the fusogenic myxovirus proteins. The mechanism of MHV-induced cell fusion may involve other, as yet unidentified domains on the S protein. Experiments to investigate these interactions at the molecular level can now be undertaken.

## Conclusion

The cloning and sequencing of the MHV-JHM structural protein genes is an important step in the study of the molecular biology of this virus and especially in the interaction between the virus, the host cell and the immune system during infection.

*Acknowledgements.* We would like to thank Helga Kriesinger for typing the manuscript, Professor V. ter Meulen for his continuous support and Roger Staden for computer programs.

## References

Armstrong J, Smeekins S, Spaan W, Rottier P, van der Zeijst B (1984a) Cloning and sequencing the nucleocapsid and E1 genes of a coronavirus. Adv Exp Med Biol 173: 155–162

Armstrong J, Niemann H, Smeekins S, Rottier P, Warren G (1984b) Sequence and topology of a model intracellular membrane protein, E1 glycoprotein, from a coronavirus. Nature 308: 751–752

Boursnell MEG, Brown TDK, Binns MM (1984) Sequence of the membrane protein gene from avian coronavirus IBV. Virus Res 1: 303–313

Boursnell MEG, Binns MM, Foulds IJ, Brown TDK (1985) Sequences of the nucleocapsid genes from two strains of avian infectious bronchitis virus. J Gen Virol 66: 573–580

Collins AR, Knobler RL, Powell H, Buchmeier MJ (1982) Monoclonal antibodies to murine hepatitis 4 (strain JHM) define the viral glycoprotein responsible for attachment and cell-cell fusion. Virology 119: 358–371

Frana MF, Behnke JN, Sturman LS, Holmes KV (1985) Proteolytic cleavage of the E2 glycoprotein of murine coronavirus: host dependent differences in proteolytic cleavage and cell fusion. J Virol 56: 912–920

Garoff H (1985) Using recombinant DNA techniques to study protein targeting in the eucaryotic cell. Annu Rev Cell Biol 1: 403–445

Holmes KV (1985) Replication of coronaviruses. In: Fields BN (ed) Virology. Raven, New York, pp 1331–1343

Knobler RL, Lampert PW, Oldstone MBA (1982) Virus persistence and recurring demyelination produced by a temperature-sensitive mutant of MHV-4. Nature 298: 279–280

Kyte J, Doolittle RF (1982) A simple method for displaying the hydropathic character of a protein. J Mol Biol 157: 105–132

Lai MMC, Brayton PR, Armen RC, Patton CD, Pugh C, Stohlman SA (1981) Mouse hepatitis virus A59: mRNA structure and genetic localization of the sequence divergence from hepatotropic strain MHV-3. J Virol 39: 823–834

Lai MMC, Baric RS, Brayton PR, Stohlman SA (1984) Characterization of leader RNA sequences on the virion and mRNAs of mouse hepatitis virus, a cytoplasmic RNA virus. Proc Natl Acad Sci USA 81: 3626–3630

Massa P, Dörries R, ter Meulen V (1986) Virus particles induce Ia antigen expression on astrocytes. Nature 320: 543–546
Neuberger A, Gottschalk A, Marshall RD, Spiro RG (1972) Carbohydrate peptide linkages in glycoproteins and methods for their elucidation. In: Gottschalk A (ed) Glycoproteins: their composition, structure and function. Elsevier, Amsterdam, pp 450–490
Niemann H, Klenk HD (1981) Coronavirus glycoprotein E1, a new type of viral glycoprotein. J Mol Biol 153: 993–1010
Pfleiderer M, Skinner MA, Siddell SG (1986) Coronavirus MHV-JHM: nucleotide sequence of the mRNA that encodes the membrane protein. Nucleic Acid Res 14: 6338
Rose JK, Adams GA, Gallione CJ (1984) The presence of cysteine in the cytoplasmic domain of the vesicular stomatis virus glycoprotein is required for palmitate addition. Proc Natl Acad Sci USA 81: 2050–2054
Rottier PJM, Brandenburg D, Armstrong J, van der Zeijst BAM, Warren G (1984) Assembly in vitro of a spanning membrane protein of the endoplasmic reticulum. The E1 glycoprotein of coronavirus mouse hepatitis virus A59. Proc Natl Acad Sci USA 81: 1421–1425
Schmidt I, Skinner MA, Siddell SG (1987) Nucleotide sequence of the gene encoding the surface projection glycoprotein of coronavirus MHV-JHM. J Gen Virol 68: 47–56
Schmidt MFG (1982) Acylation of viral spike glycoproteins: a feature of enveloped RNA viruses. Virology 116: 327–338
Siddell SG (1982) Coronavirus JHM: tryptic peptide fingerprinting of intracellular polypeptides and virion protein. J Gen Virol 62: 259–269
Siddell SG (1983) Coronavirus JHM: coding assignments of subgenomic mRNAs. J Gen Virol 64: 113–125
Siddell SG (1987) Organization and expression of coronavirus genomes. In: Rowlands DJ, Mayo MA, Mahy BWJ (eds) The molecular biology of the positive-strand RNA viruses. Academic, London, pp 117–127
Siddell SG, Wege H, Barthel A, ter Meulen V (1981) Coronavirus JHM: a virion associated protein kinase. J Gen Virol 52: 235–243
Siddell SG, Anderson R, Cavanagh D, Fujiwara K, Klenk HD, Macnaughton MR, Pensaert M, Stohlman SA, Sturman L, van der Zeijst BAM (1983a) Coronaviridae. Intervirology 20: 181–189
Siddell SG, Wege H, ter Meulen V (1983b) The biology of coronaviruses. J Gen Virol 64: 761–776
Skinner MA, Siddell SG (1983) Coronavirus JHM: nucleotide sequence of the mRNA that encodes nucleocapsid protein. Nucleic Acids Res 11: 5045–5054
Staden R (1982) An interactive graphics program for aligning nucleic acid and aminoacid sequences. Nucleic Acids Res 10: 2951–2961
Staden R (1986) The current status and portability of our sequence handling software. Nucleic Acids Res 14: 217–231
Sturman LS, Holmes KV (1983) The molecular biology of coronaviruses. Adv Virus Res 28: 35–112
Sturman LS, Ricard CS, Holmes KV (1985) Proteolytic cleavage of the E2 glycoprotein of murine coronavirus: activation of cell-fusing activity of virions by trypsin and separation of two different 90K cleavage fragments. J Virol 56: 904–911
Watanabe R, Wege H, ter Meulen V (1983) Adoptive transfer of EAE-like lesions from rats with coronavirus-induced demyelinating encephalomyelitis. Nature (Lond) 305: 150–153
White J, Kielian M, Helenius A (1983) Membrane fusion proteins of enveloped animal viruses. Rev Biophys 16 (2): 151–195

# Molecular Cloning of Flavivirus Genomes for Comparative Analysis and Expression*

CHARLES M. RICE[1], LYNN DALGARNO[2], RICARDO GALLER[3], YOUNG S. HAHN, ELLEN G. STRAUSS, and JAMES H. STRAUSS

Division of Biology, California Institute of Technology, Pasadena, CA 91125, USA

## Introduction

The flaviviruses, recently reclassified as the only genus of the family Flaviviridae, constitute a large group of enveloped viruses that cause widespread human and animal disease (Shope 1980). Most flaviviruses are transmitted in nature to vertebrate hosts via arthropod vectors (usually ticks or mosquitos) although some appear not to utilize invertebrate hosts in their life cycle (Chamberlain 1980).

In an effort to place the molecular biology of the flaviviruses on a solid footing, we have undertaken the complete sequence of the type flavivirus, yellow fever, followed by comparative sequence analysis of distantly related members of the genus. Such comparative data allow the identification of potentially important features of RNA and protein structure by virtue of their conservation during evolution (Strauss and Strauss 1983). As systems are designed for the expression of viral genes and for the expression of infectious or replication-competent viral transcripts from cloned cDNA, the importance of such features becomes amenable to direct experimental testing using directed mutagenesis. In this review we present methods used for such comparative analysis and some of the results obtained with three flaviviruses: yellow fever (YF), Murray Valley encephalitis (MVE), and dengue-2 (DEN2).

## Isolation of Flavivirus RNA

It is essential for obtaining large cDNA inserts of viral RNAs that the starting RNA be intact. For flaviviruses the problem is exacerbated by the low virus yields often obtained, which makes extraction of intact RNA more difficult. For YF virus purifi-

---

* This work was supported in part by Grants AI 20612 and AI 10793 from NIH and Grant PCM 8316856 from NSF, and by contract DAMD 17-85-C-5223 from the U.S. Army Medical Research and Development Command. R.G. was supported by a fellowship from CNPq.

[1] *Present address:* Department of Microbiology and Immunology, Washington University School of Medicine, Box 8093, 660 South Euclid Ave., St. Louis, MO 63110-1093, USA

[2] *Present address:* Biochemistry Department, The Faculties, Australian National University, P.O. Box 4, Canberra, Australia 2601

[3] *Present address:* Depart. Bioquimica e Biologia Molecular, Fundacáo Oswaldo Cruz, Av. Brasil 4365, Manguinhos, Rio de Janeiro, R.J. CEP 21040, Brazil

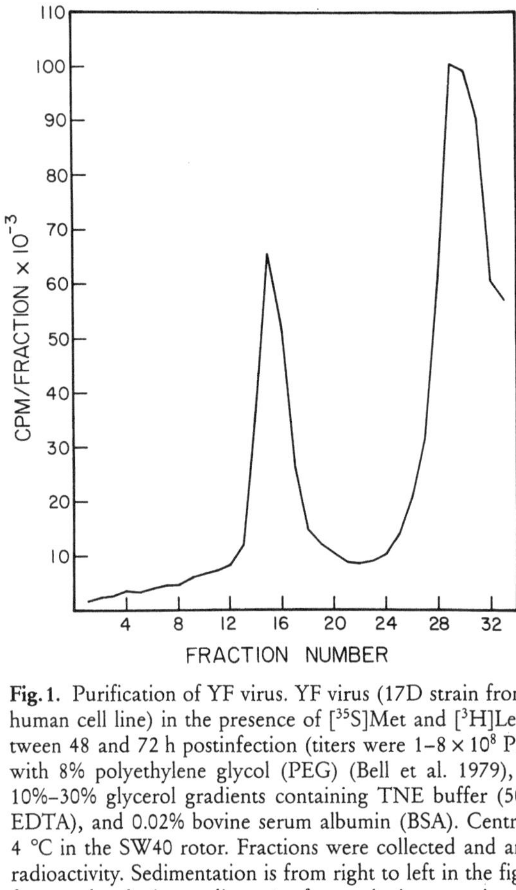


**Fig. 1.** Purification of YF virus. YF virus (17D strain from the ATCC) was grown in SW13 cells (a human cell line) in the presence of [$^{35}$S]Met and [$^{3}$H]Leu. Virus in the medium was harvested between 48 and 72 h postinfection (titers were 1–8 × $10^8$ PFU/ml) and concentrated by precipitation with 8% polyethylene glycol (PEG) (Bell et al. 1979), resuspended in buffer, and layered over 10%–30% glycerol gradients containing TNE buffer (50 m*M* NaCl, 100 m*M* Tris pH 8.6, 1 m*M* EDTA), and 0.02% bovine serum albumin (BSA). Centrifugation was for 1.5 h at 36000 rpm and 4 °C in the SW40 rotor. Fractions were collected and an aliquot of each fraction was assayed for radioactivity. Sedimentation is from right to left in the figure. If RNA was to be prepared, the virus from such velocity gradients (or from velocity potassium tartrate gradients) was further purified by isopyknic banding: pooled peaks were diluted with buffer if necessary and layered over linear potassium tartrate gradients (heavy solution: 45% w/v potassium tartrate, 30% v/v glycerol, 100 m*M* Tris-Cl pH 8.5, 1 m*M* EDTA, 200 μg/ml BSA; light solution: 30% v/v glycerol, 50 m*M* NaCl, 100 m*M* Tris-Cl pH 8.5, 1 m*M* EDTA, 200 μg/ml BSA) (Roehrig et al. 1982) and centrifuged to equilibrium (e.g., overnight in SW40 rotor at 30000 rpm at 4 °C)

cation we have found that the use of glycerol and potassium tartrate gradients for sedimentation velocity and for buoyant-density centrifugations result in highly purified virions (Fig. 1) (use of sucrose in gradients led to loss of infectivity and disrupted virus particles). From such virus preparations it is possible to obtain high molecular weight flavivirus RNA (Fig. 2). Although such RNA preparations are sometimes contaminated with low molecular weight RNAs, presumably rRNA derived from lysed host cells, or with what appears to be mitochondrial DNA (a sharp band that does not label with [$^{3}$H]uridine) (Fig. 2), such preparations are suitable for molecular cloning. Alternatively, we have found it possible to purify the RNA on sucrose gradients or by electrophoresis in low-gelling temperature (LMT) agarose and/or to treat it with RNase-free DNase I prior to cDNA synthesis.

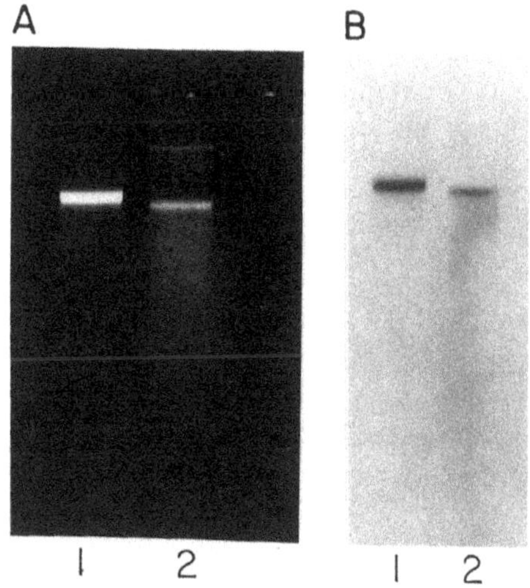


**Fig. 2 A, B.** Preparation of YF virus RNA. YF virus, labeled with [$^3$H]uridine, was purified, as described in Fig. 1. The pooled virus peak was diluted with TNE buffer and pelleted through a 15% w/v sucrose cushion in TNE (15% glycerol rather than 15% sucrose is better for preserving the integrity of the virion) in an SW 50.1 rotor at 40000 rpm at 4 °C for 4 h. The supernatant was removed completely and the pellet was resuspended in TNE containing 1% sodium dodecyl sulfate (SDS) and extracted immediately with phenol and chloroform as previously described (Hsu et al. 1973). RNA was precipitated twice with ethanol, washed once with 80% ethanol, and resuspended to 0.5–1 µg/µl in 10 m*M* TrisCl pH 8.0. Typical yields ranged from 1–2 µg of virion RNA from $10^8$ cells. An aliquot of the RNA preparation was fractionated by electrophoresis in a 1.2% agarose gel in E buffer (40 m*M* Tris base, 37 m*M* sodium acetate buffer, 1 m*M* EDTA, pH 7.8) and 0.1% SDS, and visualized either by ethidium bromide staining and UV illumination (**A**) or by fluorography (**B**). *Lane 1* contains Sindbis virus 49S virion RNA (11.7 kb) as a marker and *lane 2* is YF RNA

## Synthesis of cDNA

Initial libraries were made using random priming, since this should give a fairly complete representation of the genome and since the enzymatic modification of the 3′ terminus of intact YF and some other flavivirus RNAs has proven inefficient. Yields for first strand synthesis varied from 20% to 50% (expressed as the proportion of RNA template converted to cDNA); 50%–60% of the first strand was typically converted to second strand. Preparations of YF cDNA are illustrated in Fig. 3, assayed by alkaline agarose gel electrophoresis (McDonnell et al. 1977). Because the size distribution of first and second strands was virtually identical, showing that the two strands were not covalently linked, there was no need to use nuclease S1 to remove hairpins (Fig. 3 A); (Gubler and Hoffmann 1983). After methylation of double-stranded (ds) cDNA with *Eco*RI methylase (Fig. 3 B), the 3′→5′ exonuclease and polymerase activities of T4 DNA polymerase were used to increase the yield of flush ends. Phosphorylated *Eco*RI linkers were then ligated to the ends of the ds cDNA with T4 DNA ligase. After digestion with an excess of *Eco*RI (this enzyme was chosen because it is available in concentrated stocks and is relatively free of contaminating nonspecific nucleases), the ds cDNA with *Eco*RI cohesive ends was

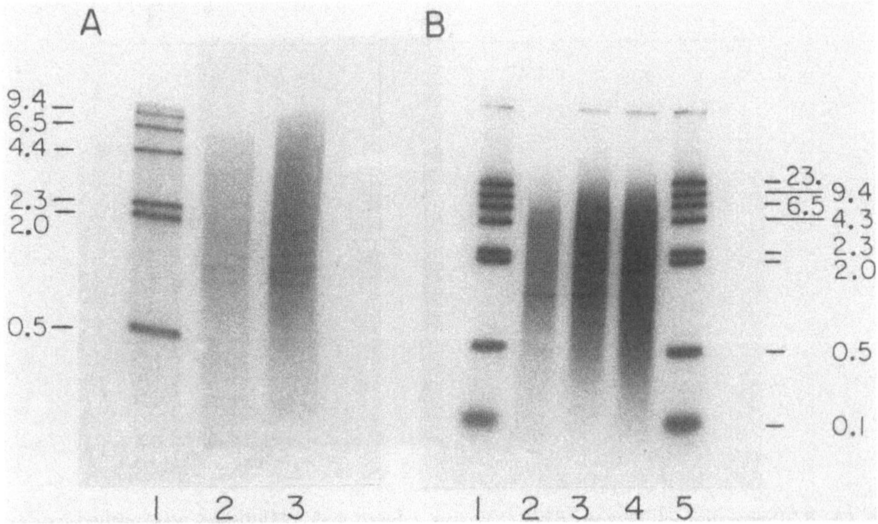


**Fig. 3 A, B.** cDNA synthesis. **A** For first strand cDNA synthesis RNA was heated to 90 °C for 2 min and chilled quickly on ice. The reaction volume of 20–30 μl contained 50 m*M* Tris-Cl pH 8.3 (at 42 °C), 8 m*M* $MgCl_2$, 50 m*M* KCl, 1 m*M* of each of the four dNTPs, 10 m*M* dithiothreitol (DTT), 40 μg/ml actinomycin D, 100 units/ml RNasin, 20–100 μg/ml template RNA, 10–20 units AMV RDDP/μg RNA, and DNase I degraded calf thymus DNA (80 μg/ml) as primer. After incubation at 42.5 °C for 30–60 min, EDTA and SDS were added to 25 m*M* and to 0.25% respectively, and the mixture was extracted with an equal volume of phenol: chloroform: isoamyl alcohol (49.5: 49.5: 1) saturated with 0.1 *M* Tris-Cl pH 7.5, containing either 8-hydroxyquinoline 0.08% or 2-mercaptoethanol 0.1% as an antioxidant. One half volume of 7.5 *M* ammonium acetate was added to the aqueous layer and the RNA: DNA hybrids were precipitated by the addition of three volumes of cold 100% ethanol. After freezing for 10 min on dry ice, the sample was warmed to 0 °C and the precipitate collected by centrifugation for 5 min in an Eppendorf microfuge. The pellet (often invisible) was resuspended in a small volume of 10 m*M* TrisCl pH 8.0, 0.1 m*M* EDTA (TE), reprecipitated as described above, washed with 80% ethanol, and air-dried or lyophilized briefly. The RNA: DNA hybrids were resuspended in a small volume (10–20 μl) of TE and stored at −70 °C. Second strand cDNA synthesis was carried out essentially as described by Okayama and Berg (1982), the cDNA: RNA hybrids being used directly for second strand synthesis. The reaction contained 20–40 μg/ml cDNA: RNA hybrid, 100 m*M* HEPES pH 6.9, 4 m*M* $MgCl_2$, 70 m*M* KCl, 1.5 m*M* 2-mercaptoethanol, 100 μ*M* dNTPs, 0.15 m*M* βNAD, 50 units/ml *E. coli* ligase (New England Biolabs), 600 units/ml *E. coli* DNA polymerase I (New England Biolabs), and 10 units/ml *E. coli* RNase H (Bethesda Research Laboratories). Incubation was for 1 h at 12 °C followed by 1 h at 20 °C. The reaction was terminated by the addition of EDTA and SDS; 20 μg carrier yeast tRNA were then added and the mixture was extracted, precipitated, washed, and resuspended as described for first strand synthesis. Samples of first strand cDNA (lane 2) or second strand cDNA (lane 3) were separated in an alkaline 1% agarose gel. The markers in lane 1 are end-labeled fragments of λ DNA digested with *Hind* III; the sizes are indicated in kilobases. **B** The randomly primed samples of duplex cDNA (ds cDNA) were methylated using *Eco* RI methylase. Reactions contained 30 μg/ml ds cDNA, 100 m*M* Tris-Cl pH 7.5, 10 m*M* EDTA, 6 μ*M* *S*-adenosylmethionine, 300 units/ml *Eco* RI methylase (New England Biolabs), and were incubated at 37 °C for 1 h, followed by extraction, precipitation, and washing as described above. The ds cDNA was resuspended and flush-ended using T4 DNA polymerase. Reaction contained 20–40 μg/ml DNA, 250 μ*M* dNTPs, 33 m*M* Tris-Ac pH 7.9, 66 m*M* KAc, 10 m*M* $MgCl_2$, 100 μg/ml BSA, 0.5 m*M* DTT, 1.25 units T4 DNA polymerase (New England Biolabs) per μg DNA, and was incubated at 37 °C for 30 min, followed by extraction, precipitation, and washing as described above.

separated from the digested linkers by electrophoresis in submerged minigels of LMT agarose (0.8%). Size fractions were isolated by phenol extraction after localization with long-wavelength UV and ethidium bromide staining (Maniatis et al. 1982), and ethanol precipitated with 25 μg carrier tRNA. Although it reduces the yield (typically 30%–50% of the material loaded on the gel was recovered in the 0.4- to 12-kb ds cDNA size classes) and perhaps the cloning efficiency of the ds cDNA, the agarose gel size selection allows for enrichment of longer inserts in the library, which are more generally useful for expression and reconstruction experiments than shorter inserts.

## Cloning of Flavivirus cDNA

The ds cDNA (4–20 ng) was ligated to *Eco*RI-linearized, alkaline phosphatase-treated pMT21 (four- to eightfold vector excess), and used to transform competent (Dagart and Ehrlich 1979) *E. coli* strain MC1061.1 ($r^-m^+$, probably recA56; Casadaban and Cohen 1980). pMT21 is a 1.9-kb plasmid derived from pBR322 that contains an ampicillin resistance marker and a polylinker. Approximately $6 \times 10^5$ colonies/μg of insert DNA were obtained for ds cDNA in the 0.6–2 kb, 2–4 kb, and 4–6 kb size ranges, with five- to tenfold fewer transformants with the 6–8 kb and 8–14 kb size classes. A similar method for ds cDNA synthesis and cloning has been described by Gubler and Hoffman (1983). Random colonies from each of these size classes were screened by hybridization (Grunstein and Hogness 1975) to a cDNA probe made by random priming (for YF, the RNA; had been exhaustively digested with RNase-free DNase I prior to cDNA synthesis) or simply on the basis of plasmid size and sequence analysis (MVE and DEN). In any case, plasmid insert sizes were estimated using quick alkaline-SDS lysis of colonies and electrophoresis on gels with appropriate supercoiled standards (Maniatis et al. 1982).

## Sequence of YF RNA

Our initial sequencing efforts were with YF RNA. With no preliminary information as to genome organization or sequence data with which to compare YF RNA, we first isolated a number of clones that tested positive by colony hybridization and

**Fig. 3 A, B** (continued). *Eco*RI linkers (dCGGA-ATTCCG; New England Biolabs), phosphorylated with T4 polynucleotide kinase (New England Biolabs) and ATP (Meyerowitz et al. 1980), were ligated to the ds cDNA in a small reaction volume (5–10 μl) using T4 DNA ligase ($4 \times 10^4$ units/ml as defined by New England Biolabs) in 100 m*M* Tris-Cl pH 7.5, 10 m*M* $MgCl_2$, 20 m*M* DTT, 1 m*M* ATP at 15 °C for 12–20 h. The linkers were present at a final concentration of 1–2 μ*M*. The reaction was terminated by incubation at 70 °C for 10 min, diluted to 50 μl in *Eco*RI restriction buffer, and digested with an excess of *Eco*RI (typically about 150–200 units for 4 h at 37 °C). Samples of ds cDNA after *Eco*RI methylation (lane 2), addition of *Eco*RI linkers (lane 3), or after digestion with *Eco*RI (lane 4) were separated in a 1% agarose non-denaturing gel. End-labeled lambda *Hind*III markers are shown in lanes 1 and 5

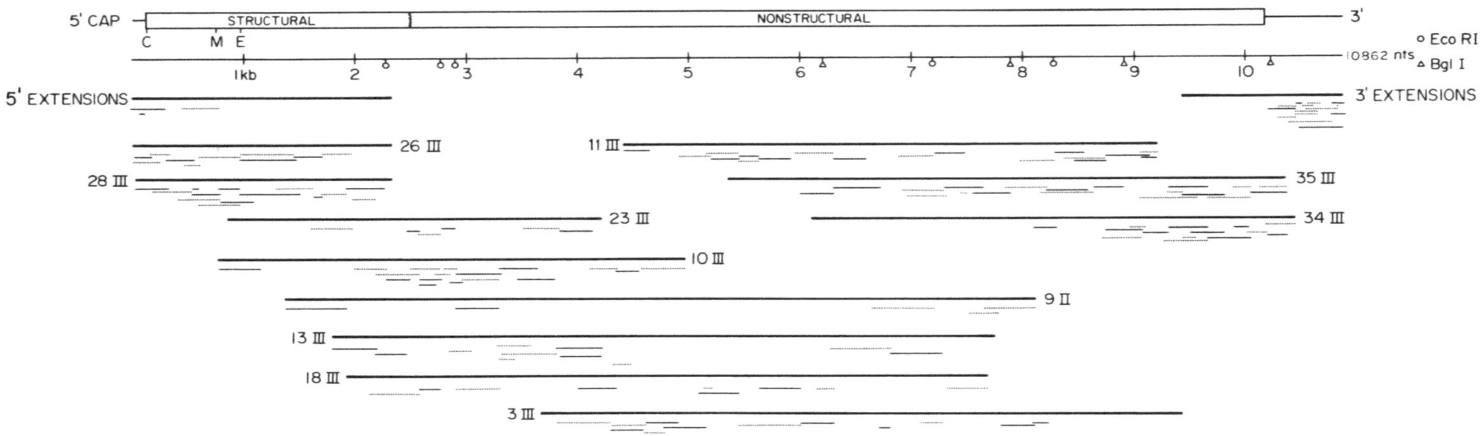


**Fig. 4.** Sequencing strategy used to obtain the entire YF sequence from cDNA clones. A representation of the YF genome is shown above as scale bar with nucleotides numbered in kilobases (*circles* and *triangles* represent *Eco*RI and *Bgl*I restriction sites respectively). Below are shown various cDNA clones drawn to scale, including clones obtained by primer extension at the 5′ end or by poly(A) tailing of the 3′ end followed by oligo(dT) priming of cDNA synthesis (see Fig. 5). Regions sequenced (by chemical sequencing) are indicated below each clone representation. For chemical sequencing restriction fragments with 5′ protruding ends were end-labeled using the Klenow fragment of *E. coli* DNA polymerase I, digested with a second restriction enzyme, and the resulting end-labeled fragments separated on preparative bisacrylcystamine cross-linked polyacrylamide gels (Garfinkel et al. 1983). The recovered fragments were sequenced by the chemical sequencing method (Maxam and Gilbert 1980) as modified by Smith and Calvo (1980), using five reactions. DNA sequences on the strand corresponding to the YF genome are indicated by *dotted lines*, and *solid lines* represent sequences determined from the complementary strand

performed preliminary restriction mapping, using primarily enzymes with six base recognition sequences. Because only 38% of the clones from the largest size class of cDNA used (8–14 kb) were positive, insert sizes of 2.3–7 kb wcrc uscd for thc rc-striction mapping, and the restriction map suggested that clones containing > 10 kb of YF RNA sequence had been isolated.

After this preliminary characterization, a sequencing strategy was devised to obtain the total sequence (Fig. 4). Restriction fragments were 3′ end-labeled and sequenced using the base-specific chemical cleavage method (Maxam and Gilbert 1980). The sequence was determined for both strands and on at least two independent cDNA clones; this allows the detection of heterogeneity in the cloned cDNA population (either due to heterogeneity in the RNA templates or random errors introduced during reverse transcription and subsequent cloning) and the sequence which has been reported (Rice et al. 1985) is a consensus of these sequences. The actual number of verifiable differences between clones was on the order of $1/10^4$. The origin(s) of these changes is unclear, since this figure is within the range of the error frequencies which have been estimated for reverse transcriptase (Battula and Loeb 1974) and also near the level of heterogeneity expected in an RNA virus population (Domingo et al. 1978; Reanny 1982; Holland et al. 1982). In either case, when using these clones to reconstruct segments of the YF genome for expression studies, we have preferentially used cloned regions containing the consensus sequence or, when this would be difficult, regions in which only silent changes (with respect to protein coding) are found.

The sequence obtained from these clones totaled 10330 nucleotides and was completely overlapped (Fig. 4). This represents 95% of the total RNA sequence, with only 12–15 nucleotides at the 5′ end and about 520 nucleotides at the 3′ end missing. As previously reported there is a long open reading frame in the RNA (Rice et al. 1985), and comparison of this (deduced) protein sequence with the N-terminal amino acid sequence data from several YF proteins allowed their localization within the RNA sequence. Thus the construction and analysis of a random library generated most of the genome sequence.

## Cloning the Extreme 5′ and 3′ Termini of the YF Genome

Once the sequence of the major part of the YF genome had been determined from overlapping clones, we were able to devise a directed strategy for obtaining clones representing the remaining portions of the genome (Fig. 5). For cloning the extreme 5′ end the 1066-nucleotide *Sau*3AI fragment from YF clone $28_{III}$ (see Fig. 4) was gel-purified and used to prime cDNA synthesis on YF RNA. After first strand synthesis, the cDNA was C-tailed using terminal deoxynucleotidyl transferase, and oligo$(dG)_{12-18}$ (12.5 μg/ml) was used to prime second strand synthesis. This ds cDNA was cleaved with *Ava I* and the extended product gel-purified and cloned into $28_{III}$ which had been digested with *Sma I* (which cuts in the polylinker of pMT21) followed by *Ava I* (which should generate clones containing the 5′ 2.3 kb of the YF genome). The resulting clones were screened by restriction and sequence analysis. Two of the clones contained the same extended sequence followed by a stretch of oli-

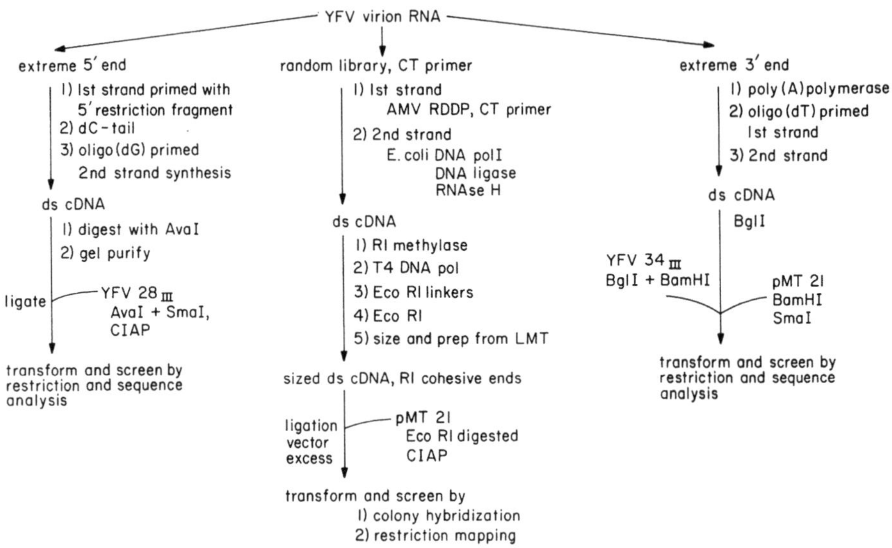


**Fig. 5.** Schematic diagram of the methods used to obtain cDNA libraries that represent the entire YF sequence. Some techniques have been described in the legends to other figures. For polyadenylation of YF RNA, virion RNA (4 μg) was heated to 90 °C for 2 min, chilled on ice, and polyadenylated using *E. coli* poly(A) polymerase (Arias et al. 1984). The reaction contained (in a final volume of 30 μl) 50 m*M* Tris-Cl pH 7.9, 10 m*M* $MgCl_2$, 2.5 m*M* $MnCl_2$, 250 m*M* NaCl, 0.2 m*M* ATP, 500 μg/ml BSA, 10 m*M* DTT, 30 units RNasin (Promega Biotec), and 0.12 units *E. coli* poly(A) polymerase (from C. Arias). Incubation was at 37 °C for 1 h and the reaction was terminated by addition of EDTA to 25 m*M* followed by phenol: chloroform extraction and ethanol precipitation. RNA was resuspended at 0.5–1 μg/μl in 10 m*M* Tris-Cl pH 8.0 and stored at −70 °C. Under these conditions an average of 10–20 A residues were added per molecule of virion RNA

go(dG). The length of this extended sequence agreed (within two to three nucleotides) with the size of the longest primer extension product that had been observed before cloning. In addition, this sequence is homologous with the 5′ genomic RNA sequence determined for the flavivirus West Nile (WN) (Wengler and Wengler 1981), and thus probably corresponds to the extreme 5′ end of the YF genome.

The 3′-enriched library was constructed using YF RNA which had been polyadenylated with *E. coli* poly(A) polymerase. Oligo$(dT)_{12-18}$ (12.5 μg/ml) was used to prime first strand synthesis. After second strand synthesis, the ds cDNA was cleaved with *Bgl*I (which potentially generates many cohesive 3′-terminal overhangs) and cloned using *Bgl*II- and *Sma*I-digested pMT21 and a *Bam*HI-*Bgl*I fragment of YF clone $35_{III}$. In this library only fragments in the ds cDNA preparation which had the same *Bgl*I cohesive end as the fragment from $35_{III}$ (which was the 3′-terminal *Bgl*I site in the YF genome; see Fig. 4) should be cloned. This situation arises because the recognition sequence of *Bgl*I is 5′-GCCNNNN↓NGGC-3′, where cleavage occurs at the point marked with the arrow and where N may be any nucleotide. Thus most *Bgl*I cleavage sites are different, and the *Bgl*I site in fragment $35_{III}$, GCCACCT↓GGGC, results in a cohesive end that will selectively base-pair with the complementary cohesive end produced by cleavage at the same *Bgl*I site (by random chance only 1/256 *Bgl*I sites would be expected to generate the same sequence).

This selection step to obtain the correct 3′ end clone was used because polyadenylation of the 3′ end of YF is inefficient and we wanted to select against ds cDNA that arose from sources other than polyadenylation of the 3′ terminus of YF RNA. The resulting clones were screened by restriction and sequence analysis. Six independent clones were characterized and each contained a poly(dA) stretch followed by the same sequence, which was homologous with the 3′-terminal RNA sequence of WN virus (Wengler and Wengler 1981). Several of the poly(dA) tracts were longer than 20 nucleotides (therefore longer than the length of the oligo[dT] primers used for cDNA synthesis), ruling out the possibility that these clones originated by selective internal priming at a polypurine-rich region in the YF genome. It is therefore likely that these clones correspond to the extreme 3′ end of the YF genome. The poly(dA) tract presumably originated in the poly(A) polymerase reaction but we cannot exclude the possibility that a small fraction of the YF genomic RNA contains poly(A). Deubel et al. (1983) have shown that most YF RNA does not bind to oligo(dT) cellulose, and also that the 3′-terminal nucleotide is U, which agrees with the 3′-terminal YF sequence deduced from cloned cDNA (Rice et al. 1985; Grange et al. 1985).

## Cloning of Other Flaviviruses

Once the sequence of YF RNA was known, it was possible to obtain libraries of other flavivirus cDNAs in a rapid and directed fashion. Randomly primed libraries of MVE and DEN2 were constructed and colonies screened for insert size by agarose gel electrophoresis after alkaline-SDS colony lysis (Maniatis et al. 1982). After preliminary restriction analysis, selected clones were end-labeled at the *Eco*RI sites used for cloning or other convenient restriction sites in the vector polylinker, and their terminal sequences were determined. These sequences and their complements were translated in all three possible reading frames, and the protein sequences were compared by a dot matrix homology program to the YF polyprotein sequence (some examples with MVE are shown in Fig. 6). For most of the sequences, obvious homology with the YF sequence was detected in one of the reading frames, thus allowing the alignment of the MVE clones and DEN2 clones (Fig. 7) relative to the YF genome. Our experience has been that a sequence of 100–200 nucleotides from MVE or DEN2 could be aligned along the genome in most cases. Sequences from the regions corresponding to the C protein (Fig. 6 A) or to the 5′ and 3′ noncoding regions (Fig. 6 D) were more difficult to align, however, due to generally lower levels of colinear protein or nucleotide sequence homology (see also Dalgarno et al. 1986). This method of alignment (without extensive restriction analysis or cross-hybridization studies) is rapid and should be applicable to any flavivirus. Considerable nucleotide sequence data are also obtained in the process, and in addition clones whose end sequences and measured insert size do not conform and thus contain sequence rearrangements of some form are readily apparent.

Because of the degeneracy in the genetic code and the fact that flaviviruses (Dalgarno et al. 1986) and other virus families such as alphaviruses (Strauss and Strauss 1986) have randomized the codon used for any conserved amino acid in the protein sequence, it is more efficient to search for such homologies (for alignment purposes)

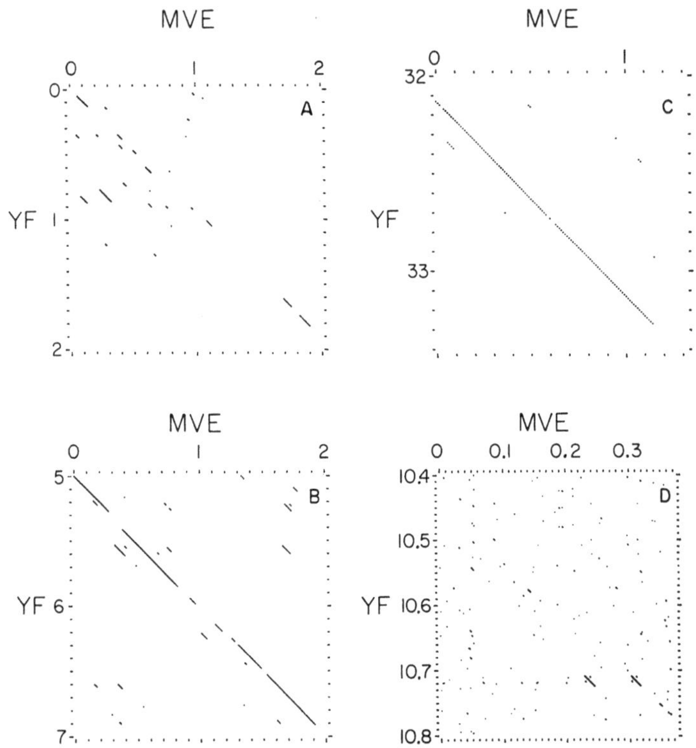


**Fig. 6 A–D.** Alignment of MVE clones with the YF sequence. Nucleotide sequences indicated as open boxes in Fig. 7 were translated in all six possible reading frames and the amino acid sequences compared with the YF polyprotein sequence. Dot matrices between the correct open reading frame in the MVE sequence of clones 2/1/22, 1/1/12, and 2/2/6 and the corresponding homologous region from the YF sequence are shown in **A–C** respectively. The translated sequence of the box in clone 2/2/38 showed no detectable homology with the YF polyprotein sequence and was compared at the nucleotide level with the YF 3′ untranslated region (**D**). Four of nine amino acids (**A–C**) or five of six nucleotides (**D**) were scored as a match. The YF sequences are numbered as described in Rice et al. (1985) and coordinates are given in amino acids ($\times 10^{-2}$) (**A–C**) or in kb (**D**)

at the amino acid sequence level than at the nucleotide level. An example of an amino acid sequence conserved among YF, WN, MVE, and DEN2 is shown in Fig. 8 together with the nucleotide sequence encoding it in the four viruses. Although the nine amino acids encoded are perfectly conserved, only two of the amino acids are encoded by the same codon in all four viruses. Thus hybridization probes prepared from one virus cannot be readily used to screen clones from a second virus in order to identify clones from a selected region of the genome, although in principle it might be possible to synthesize highly degenerate probes for such use.

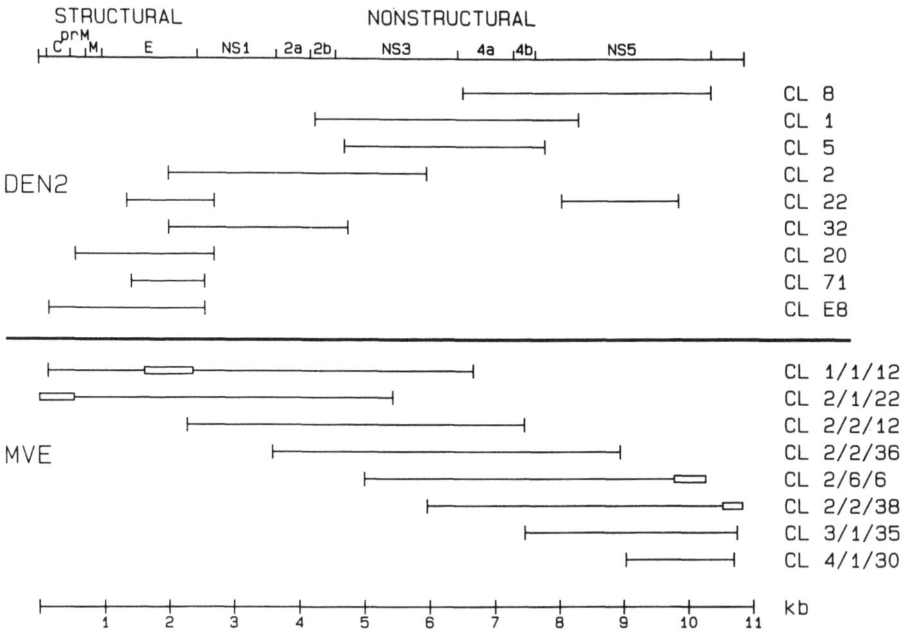


**Fig. 7.** Map of two libraries of flavivirus clones. The nine clones containing DEN2 sequences and the eight clones of MVE have been positioned along the genome by nucleotide sequence analysis of the ends of the inserts followed by computer searches as shown in Fig. 6 to identify regions homologous to the YF genome. Most DEN2 clones were found in the original screening; clone E8 was isolated after screening with an E protein-specific probe. A map of the genome of YF showing the locations of the encoded proteins is shown at the top. *Scale bar* is shown below with nucleotide numbers in kilobases. The *open boxes* in the MVE clones indicate the regions shown as dot plots in Fig. 6. The MVE stock used was a sample of the original Australian isolate (MVE/1/1951; Berge 1975) obtained from R. Shope. DEN2 RNA from the PR159 (S-1) strain was provided by Dr. Joel Dalrymple

| | | | | | | | | |
|---|---|---|---|---|---|---|---|---|
| | P | P | F | G | D | S | Y | I |
| YF | CCA | CCU | UUU | GGA | GAC | AGC | UAC | AUU |
| | P | P | F | G | D | S | Y | I |
| WN | CCC | CCG | UUU | GGU | GAC | UCU | UAC | AUC |
| | P | P | F | G | D | S | Y | I |
| MVE | CCA | CCC | UUC | GGA | GAC | UCA | UAC | AUU |
| | P | P | F | G | D | S | Y | I |
| DEN2 | CCU | CCA | UUC | GGA | GAC | AGC | UAC | AUC |
| CODONS | 3 | 4 | 2 | 2 | 1 | 3 | 1 | 2 |

**Fig. 8.** Randomization of codons in conserved amino acid sequences. Short stretches of nucleotide sequence encoding a portion of the E protein of YF, WN, MVE, and DEN2 are compared. This sequence begins at nucleotide 2078 of the YF genome (amino acid 369 of the E protein). The YF sequence is from Rice et al. (1985), WN from Wengler et al. (1985), MVE from Dalgarno et al. (1986), and DEN from unpublished data from our laboratory

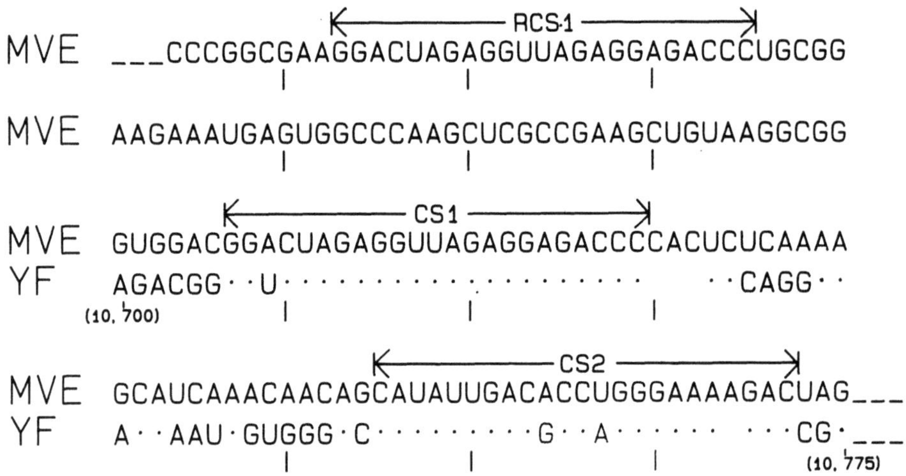


**Fig. 9.** Conserved nucleotide sequences in the 3′ noncoding region of MVE and YF. Nucleotide sequences in the 3′ noncoding regions of MVE and YF (numbered according to Rice et al. 1985) are aligned. Identical positions in the YF sequence relative to the MVE sequence are indicated by *dots*. The sequence homology becomes negligible 5′ to position 10705 in the YF nucleotide sequence. Conserved sequences *(CS1* and *CS2)* and a repeat of CS1 in MVE *(RCS1)* are indicated

## Conserved Sequences in the Flavivirus 3′ Noncoding Region

A comparison of the terminal sequence of MVE clone 2/2/38 with the 3′ noncoding region of YF revealed two regions of short but striking homology (Fig. 6 D). These sequences are aligned in Fig. 9. This MVE clone lacks the extreme 3′ terminus and begins (at its 3′ end) with a sequence that is homologous with the YF sequence starting 88 nucleotides from the 3′ terminus (Fig. 7). Twenty-two of the first 28 nucleotides of the MVE clone are identical to the YF sequence from positions 10749 to 10775, with a gap in the YF sequence. There is a second homologous region containing 22 of 23 nucleotides identical (positions 10706 to 10728 in the YF sequence) separated from the first region by a gap of 20–23 bases (Fig. 7). An identical repeat of the latter sequence, 23 nucleotides in length, is found 55 nucleotides upstream in the MVE genome, but such a repeat is not found in YF. Three closely spaced repeated sequences each 44 nucleotides in length have been found in the YF 3′ noncoding region (Rice et al. 1985) between positions 10372 and 10520 (Fig. 10), but data for the corresponding region of MVE are not yet available. Oligonucleotides complementary to these conserved sequences might be useful for isolation of flavivirus-specific RNAs or as universal primers for cDNA synthesis. A schematic diagram of these conserved and/or repeated sequences is shown in Fig. 11.

We have previously reported that the 3′-terminal 88 nucleotides of YF could form a secondary structure with a free energy calculated to be −40 to −46 kcal/mol. In this structure the 3′-terminal U of YF RNA is base-paired to nucleotide 10775. If a similar structure exists in MVE it is conceivable that clone 2/2/38, which begins with a nucleotide corresponding to nucleotide 10774 of YF, could have arisen by

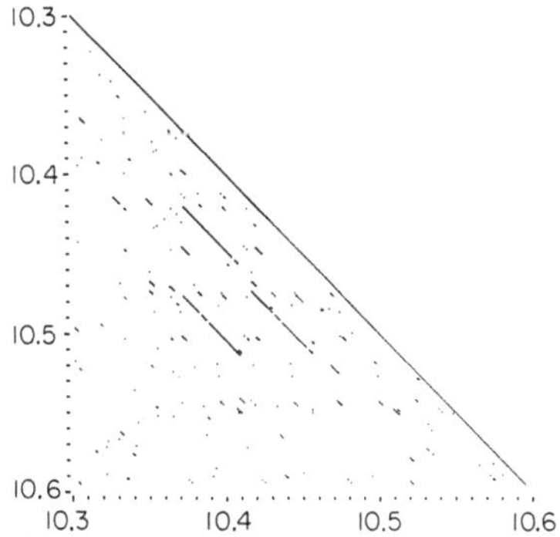


A

```
10372  AUAACCGGGAUACAAACCACGGGUGGAGAACCGGACUCCCCACA
10420  GA..........U.........C...........G....G...U
10476  GA..........A....U....A................A....
```

B

**Fig. 10 A, B.** Repeated sequences in the 3′ noncoding region of YF. **A** Dot matrix analysis of 300 nucleotides of YF RNA between nucleotides 10300 and 10600 from the 5′ end (the entire RNA is 10862 nucleotides in length). The three repeated sequences are clearly seen, and no other repeated sequences of significant length are present in the genome. In this plot an identity of five of six nucleotides was scored as a match. **B** List of the nucleotide sequences in the three repeats seen in **A.** The first *(upper)* sequence is shown in full. A *dot* in either of the *lower* two sequences indicates that the nucleotide at that position is the same as in the upper sequence. The repeated sequence element is 42 nucleotides in length and there are four to five changes between any two of the three sequences

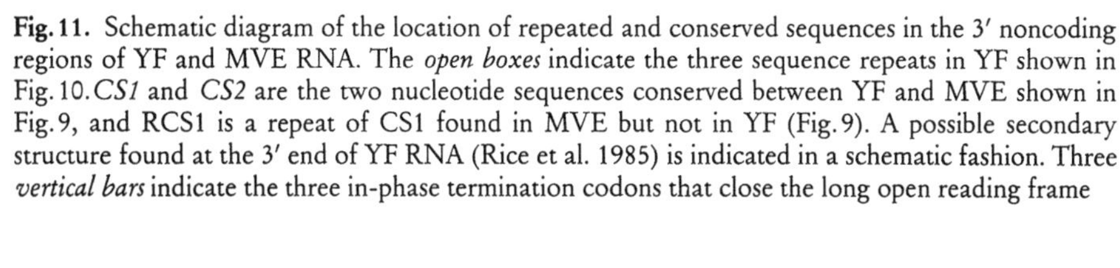


**Fig. 11.** Schematic diagram of the location of repeated and conserved sequences in the 3′ noncoding regions of YF and MVE RNA. The *open boxes* indicate the three sequence repeats in YF shown in Fig. 10. *CS1* and *CS2* are the two nucleotide sequences conserved between YF and MVE shown in Fig. 9, and RCS1 is a repeat of CS1 found in MVE but not in YF (Fig. 9). A possible secondary structure found at the 3′ end of YF RNA (Rice et al. 1985) is indicated in a schematic fashion. Three *vertical bars* indicate the three in-phase termination codons that close the long open reading frame

self-priming. Grange et al. (1985) have proposed an alternative secondary structure for the 3′ end of YF, involving 120 nucleotides at the 3′ end and having a calculated stability of −52 kcal/mol.

## Concluding Remarks

The flaviviruses are an important group of viruses causing a number of serious human illnesses worldwide. Construction of cDNA libraries that contain all or most of the genomic sequences of flaviviruses makes it possible to obtain the complete nucleotide sequence (Rice et al. 1985) or partial nucleotide sequences (Dalgarno et al.

1986) of flavivirus genomes as well as enabling expression experiments and new approaches to the development of flavivirus vaccines. Furthermore, these cloned sequences may be of use as diagnostic probes.

Comparing the nucleotide sequences and encoded amino acid sequences of flaviviruses also leads to definition of conserved and nonconserved domains and to models for the importance and function of nucleotide or amino acid domains in flavivirus replication (reviewed by Rice et al. 1986). As an example of this approach, we have identified two sequences of about 20 nucleotides in length in the 3′ noncoding region that are highly conserved between YF and MVE. One of these two sequences is found in two (noncontiguous) copies in MVE, but neither is repeated in YF (Figs. 8–10). The function of these conserved sequences is uncertain, but it seems likely that they are important in some way for RNA replication or for encapsidation of the RNA. They could serve as binding sites, for example, for the viral replicase or capsid protein. The position of one of these conserved sequences immediately adjacent to the hypothetical 3′-terminal secondary structure of YF RNA also suggests that it might be involved in initiation of RNA replication. The significance of the repetition of one of these sequences in MVE but not in YF is also unclear; YF does have repeated sequences in the 3′ untranslated region but it is unknown whether these sequences are conserved with MVE. We have also found repeated sequences (on the order of 60 nucleotides in length) in many alphaviruses, and these repeated sequences are sometimes conserved between two or more alphaviruses (Ou et al. 1982).

*Acknowledgements.* We thank E. M. Lenches for expert technical assistance, T. Hunkapiller for the many computer programs used in aligning the sequence data and producing the figures in this paper, and L. Hood for use of his computer facilities. MVE RNA was kindly supplied by Dr. Dennis Trent; DEN2 RNA was a gift from Dr. Joel Dalrymple.

## References

Arias CF, Lopez S, Bell JR, Strauss JH (1984) Primary structure of the neutralization antigen of SA11 simian rotavirus as deduced from cDNA sequence. J Virol 50: 657–661

Battula N, Loeb LA (1974) The infidelity of avian myeloblastosis virus deoxyribonucleic acid polymerase in polynucleotide replication. J Biol Chem 249: 4086–4093

Bell JR, Strauss EG, Strauss JH (1979) Purification and amino acid compositions of the structural proteins of Sindbis virus. Virology 97: 287–294

Berge TO (1975) International catalogue of arboviruses including certain other viruses of vertebrates, 2nd edn. National Institutes of Health, Bethesda, pp 504–505

Casadaban MJ, Cohen SN (1980) Analysis of gene control signals by DNA fusion and cloning in *Escherichia coli.* J Mol Biol 138: 179–207

Chamberlain RW (1980) Epidemiology of arthropod-borne togaviruses: the role of arthropods as hosts and vectors and of vertebrate hosts in the natural transmission cycle. In: Schlesinger RW (ed) The togaviruses. Academic, New York, pp 175–228

Dagert M, Ehrlich SD (1979) Prolonged incubation in calcium chloride improves the competence of *Escherichia coli* cells. Gene 6: 23–28

Dalgarno L, Trent DW, Strauss JH, Rice CM (1986) Partial nucleotide sequence of Murray Valley encephalitis virus: comparison of the encoded polypeptides with yellow fever virus structural and nonstructural proteins. J Mol Biol 187: 309–323

Deubel V, Crouset J, Bénichou D, Digoutte J-P, Bouloy M, Girard M (1983) Preliminary characterization of the ribonucleic acid of yellow fever virus. Ann Virol 134E: 581–588

Domingo E, Sabo D, Taniguchi T, Weissmann C (1978) Nucleotide sequence heterogeneity of an RNA phage population. Cell 13: 735–744

Garfinkel MD, Pruitt RE, Meyerowitz EM (1983) DNA sequences, gene regulation and modular protein evolution in the *Drosophila* 68C gene cluster. J Mol Biol 168: 765–789

Grange T, Bouloy M, Girard M (1985) Stable secondary structure at the 3′-end of the genome of yellow fever virus (17D vaccine strain). FEBS Lett 188: 159–163

Grunstein M, Hogness D (1975) Colony hybridization: a method for the isolation of cloned DNAs that contain a specific gene. Proc Natl Acad Sci USA 72: 3961–3965

Gubler U, Hoffman BJ (1983) A simple and very efficient method for generating cDNA libraries. Gene 25: 263–269

Holland J, Spindler K, Horodyski G, Grabau E, Nichol S, Vandepol S (1982) Rapid evolution of RNA genomes. Science 215: 1577–1585

Hsu MT, Kung HJ, Davidson N (1973) An electron microscope study of Sindbis virus RNA. Cold Spring Harbor Symp Quant Biol 38: 943–950

Maniatis T, Fritsch EF, Sambrook J (1982) Molecular cloning. Cold Spring Harbor Laboratory, Cold Spring Harbor, New York

Maxam AM, Gilbert W (1980) Sequencing end-labeled DNA with base-specific chemical cleavage. Methods Enzymol 65: 499–560

McDonnell MW, Simon MN, Studier FW (1977) Analysis of restriction fragments of T7 DNA and determination of molecular weights by electrophoresis in neutral and alkaline gels. J Mol Biol 110: 119–146

Meyerowitz EM, Guild GM, Prestidge LS, Hogness DS (1980) A new cosmid vector and its use. Gene 11: 271–282

Okayama H, Berg P (1982) High efficiency cloning of full length cDNA. Mol Cell Biol 2: 161–170

Ou J-H, Trent DW, Strauss JH (1982) The 3′-non-coding regions of alphavirus RNAs contain repeating sequences. J Mol Biol 156: 719–730

Reanney DC (1982) The evolution of RNA viruses. Annu Rev Microbiol 36: 47–73

Rice CM, Lenches EM, Eddy SR, Shin SJ, Sheets RL, Strauss JH (1985) Nucleotide sequence of yellow fever virus: implications for flavivirus gene expression and evolution. Science 229: 726–733

Rice CM, Strauss EG, Strauss JH (1986) Structure of the flavivirus genome. In: Schlesinger S, Schlesinger M (eds) The togaviridae and flaviviridae. Plenum, New York, pp 279–326

Roehrig JT, Day JW, Kinney RM (1982) Antigenic analysis of the surface glycoproteins of a Venezulean equine encephalomyelitis virus (TC-83) using monoclonal antibodies. Virology 118: 269–278

Shope RE (1980) Medical significance of togaviruses: an overview of diseases caused by togaviruses in man and in domestic and wild vertebrate animals. In: Schlesinger RW (ed) The togaviruses. Academic, New York, pp 47–82

Smith DR, Calvo JM (1980) Nucleotide sequence of the *E. coli* gene coding for dihydrofolate reductase. Nucleic Acids Res 8: 2255–2274

Strauss EG, Strauss JH (1983) Replication strategies of the single stranded RNA viruses of eukaryotes. Curr Top Microbiol Immunol 105: 1–98

Strauss EG, Strauss JH (1986) Structure and replication of the alphavirus genome. In: Schlesinger S, Schlesinger M (eds) The togaviridae and flaviviridae. Plenum, New York, pp 35–90

Wengler G, Wengler G (1981) Terminal sequences of the genome and replicative-form RNA of the flavivirus West Nile virus: absence of poly(A) and possible role in RNA replication. Virology 133: 544–555

Wengler G, Castle E, Leidner U, Nowak T, Wengler G (1985) Sequence analysis of the membrane protein V3 of the flavivirus West Nile virus and of its gene. Virology 147: 264–274

# The Ambisense Coding Strategy of the S RNA of Arenaviruses

DAVID H. L. BISHOP

NERC Institute of Virology, Mansfield Road, Oxford OX1 3SR, UK

## Introduction

Viruses are classified on the basis of the type of their genetic information (RNA or DNA), its form (single-stranded or double-stranded), and the procedures employed for virus replication (Matthews 1982). A variety of strategies are used by viruses to replicate their genomes in permissive cells. For the single-stranded RNA genome viruses (other than members of the Retroviridae which synthesize a DNA intermediate during the replication cycle), the strategy involves either the immediate translation of proteins from the viral RNA (positive-stranded viruses), or the initial synthesis of viral-complementary mRNA species that are then translated into proteins (negative-stranded viruses). For the positive-stranded RNA viruses, proteins are made either by complete translation of the infecting viral RNA species (e.g., picornaviruses, caliciviruses), or by partial translation of the infecting viral RNA (e.g., coronaviruses, alphaviruses), and subsequent synthesis of viral-sense subgenomic mRNA species that are made following the RNA replication process. Either way, viral-sense RNA species function as mRNA species. For all of the negative-stranded RNA viruses, the synthesis of mRNA species occurs at the onset of infection and is catalyzed by viral coded enzymes that are present in the infecting virus particles. Double-stranded RNA viruses (e.g., members of the Reoviridae or Birnaviridae) similarly use a virion polymerase to copy into mRNA one strand of each duplex of genomic RNA. The virion RNA of double-stranded RNA viruses is not translated by cellular ribosomes.

Examples of positive-stranded virus families include the Picornaviridae, Caliciviridae, Coronaviridiae, Togaviridae (alphaviruses), Flaviviridae and most of the plant virus groups that either have a single or segmented RNA genome (exceptions include the negative-stranded plant rhabdoviruses). Positive-stranded RNA viruses include among their members both enveloped viruses (togaviruses, flaviviruses and coronaviruses) and non-enveloped viruses (picornaviruses, caliciviruses and most of the plant virus groups). By contrast, the negative-stranded RNA viruses are all enveloped (Rhabdoviridae, Orthomyxoviridae, Paramyxoviridae, Bunyaviridae and Arenaviridae). Among these negative-stranded viruses are viruses with a single species of RNA (rhabdoviruses, paramyxoviruses) and viruses with seven to eight species (orthomyxoviruses), three species (bunyaviruses) or two species (arenaviruses) of RNA.

In this article, the evidence and implications of an ambisense coding strategy for the negative-stranded arenaviruses are discussed. The term "ambisense" is used to indicate that viral gene products are translated from viral-complementary as well as viral-sense mRNA sequences.

## The Arenaviruses

Members of the Arenaviridae include lymphocytic choriomeningitis (LCM, prototype virus of the family), Lassa (the etiologic agent of Lassa fever), Ippy, Mopeia, Mobala, Junin (the agent of Argentine hemorrhagic fever), Machupo (the agent of Bolivian hemorrhagic fever), Pichinde, Tacaribe, Tamiami, Latino, Amapari, Flexal and Parana viruses (Rawls and Leung 1979; Matthews 1982; Karabatsos 1985). All are transmitted in nature by rodents (*Mus, Calomys, Praomys, Akodon, Mastomys, Oryzomys, Arvicanthis, Lemniscomys, Neacomys, Thomasomys, Sigmodon* species), or, for Tacaribe virus, fruit-eating bats (*Artibeus* species) (Howard 1986). LCM, Lassa, Ippy, Mobala and Mopeia viruses are often referred to as the Old World arenaviruses (or the LCM complex viruses), the others have either been termed the New World arenaviruses, or the Tacaribe complex viruses.

In the normal non-fatal transuterine, transovarian or neonatally acquired infection of the host species, persistent lifelong infection by an arenavirus is common (Rawls and Leung 1979; Matthews 1982; Howard 1986). Viremia and/or viruria contribute to the syndrome and to the transmission cycle. The acquisition of an arenavirus by an adult rodent species usually results in a temporary acute phase of infection with virus removal in the survivors. The virulence of the virus in the adult rodent species depends on the virus, the host species and the route of inoculation. Infection within the rodent nest may involve transmission of virus to the offspring via milk, saliva and/or urine (Matthews 1982). Although virus spread to other species is uncommon, human infections in nature (LCM, Lassa, Junin, Machupo) and in the laboratory (e.g., LCM, Lassa, Flexal, Junin, Machupo, Pichinde, Tacaribe) have been documented (Karabatsos 1985). The symptoms and severities of human arenavirus diseases vary with the virus (Howard 1986).

Interest in arenaviruses has centered on the persistent nature of the normal rodent infection from both immunologic and molecular viewpoints as well as on the unique attributes of the arenavirus replication processes (Pedersen 1979; Rawls and Leung 1979; Compans and Bishop 1985). The effect of virus replication on the functions of infected cells in different organs of the host species is also a subject of study (Oldstone et al. 1985). Persistent infections are readily established in vitro, with results that are reviewed elsewhere (Rawls et al. 1981).

Molecular analyses have shown that arenaviruses are enveloped and pleomorphic (50–300 nm diameter) with glycoprotein projections on the surface (either one or two, G1, G2, depending on the virus) and two internal nucleocapsid structures (Pedersen 1979; Rawls and Leung 1979, Compans and Bishop 1985). As exemplified by Pichinde and LCM viruses the nucleocapsids consist of viral RNA (large, L, approximately $2 \times 10^6$ daltons; small, S, $1.1 \times 10^6$ daltons), associated with many copies of the $62–63 \times 10^3$ dalton nucleocapsid protein (N) and small amounts of an estimated

$180–200 \times 10^3$ dalton polypeptide (L protein) that may be a transcriptase/replicase component. It is unclear whether there are any non-structural proteins synthesized in virus-infected cells that are unrelated in primary sequence to the structural proteins. Ribosomes are usually incorporated within the envelope of the mature arenavirus particle (reviewed by Pedersen 1979). The existence of ribosomes within arenavirus particles is not surprising in view of the pleomorphic character of virions and the fact that arenaviruses do not effectively inhibit host cell macromolecular syntheses. Also, ribosomes are frequently observed at the sites of virus maturation (the cell plasma membrane). The presence of ribosomes in arenavirus particles is a characteristic feature that sets this family of viruses apart from other virus families (Matthews 1982). Experiments involving the growth of Pichinde virus in cells in which the ribosomes are temperature-sensitive *(ts)* have shown that the virion ribosomes are not essential for virus infectivity, since other cells can be infected at non-permissive temperatures with viruses possessing *ts* ribosomes (Leung and Rawls 1977).

## The Genetic Structure of Arenaviruses

Genetic experiments have shown that reassortant (recombinant) viruses can be generated among *ts* mutants or distinguishable wild-type strains of LCM virus or Pichinde virus (Vezza and Bishop 1977; Vezza et al. 1980; Kirk et al. 1980; Romanowski and Bishop 1983; Riviere et al. 1985), although no intertypic reassortants have been recovered (Romanowski and Bishop 1983). For both viruses the genetic studies have identified two recombination groups of mutants (L and S mutants) in agreement with the existence of two species of viral RNA. In addition, genetic studies with mutants of LCM virus have provided evidence for two groups of complementing mutants coded by the viral S RNA, indicating that the S species contains the information for at least two proteins (Romanowski and Bishop 1983). These observations have been substantiated by nucleic acid sequencing and other molecular analyses (see below). Furthermore, viruses that contain multiple copies of the S RNA species have been identified among virus populations, i.e., such viruses are genetically diploid (Romanowski and Bishop 1983). This finding probably reflects the pleomorphic character of arenavirus particles. Diploidy may contribute to the biological stability of arenaviruses in nature and to the experimental difficulties encountered in recovering arenavirus *ts* mutants (Vezza and Bishop 1977).

The observation that there are at least three types of viral protein (the N, L and one or two glycoprotein species), but only two viral RNA species, also supports the argument that one of the viral RNA species must code for at least two proteins. Analyses of reassortant Pichinde viruses (derived from parent viruses with distinguishable RNA sequences and proteins) have shown that the L RNA codes for the L protein and that the S RNA codes for both the viral glycoproteins and the N protein (Vezza et al. 1980; Harnish et al. 1983). The fact that the S RNA codes for N and GPC proteins presumably accounts for the complementation observed between certain S RNA mutants of LCM virus (Romanowski and Bishop 1983). Similar coding results have been obtained from analyses of reassortant LCM viruses generated from different virus strains (Riviere et al. 1985).

## The Ambisense Coding Strategy of the Arenavirus S RNA

The 5′ and 3′ ends of the S RNA species of Pichinde virus are complementary for approximately 20 residues (Auperin et al. 1982a, b; Auperin et al. 1984a, b) as are those of LCM virus S RNA (Auperin et al. 1982a; Romanowski and Bishop 1985; Romanowski et al. 1985). The 3′ end sequences are comparable not only between the two viruses but also for the L and S RNA species of both viruses. Identification of the end sequences has allowed DNA clones of the viral RNA species to be obtained using 3′ complementary oligodeoxynucleotide primers, viral RNA and reverse transcriptase.

Sequence analyses of DNA copies of Pichinde and LCM viral S RNA species have been reported (Pichinde virus S RNA: 3419 nucleotides, Auperin et al. 1984a, b; LCM virus S RNA: 3375 nucleotides, Romanowski and Bishop 1985; Romanowski et al. 1985). The data have indicated that for both viruses the $1.1 \times 10^6$ dalton viral S RNA codes for a $62–63 \times 10^3$ dalton N protein in a viral-complementary sequence corresponding to the 3′ half of the S RNA and a glycoprotein precursor to the G1 and G2 proteins (GPC, $56–57 \times 10^3$ daltons as the unglycosylated primary gene product) in a viral-sense sequence corresponding to the 5′ half of the RNA. To characterize this coding arrangement the term ambisense RNA has been coined (Auperin et al. 1984b; Romanowski et al. 1985). Northern analyses undertaken to confirm the identities of the N and GPC mRNA species have shown that for both LAC and Pichinde viruses the N mRNA species is viral-complementary in sequence and subgenomic in size. It corresponds to approximately half the length of the viral S RNA species (Auperin et al. 1984b; Romanowski et al. 1985). Thus the N mRNA species of LCM virus is approximately 1770 nucleotides long, while that of Pichinde virus is some 1800 nucleotides in length. The lack of binding of the viral N mRNA species to oligo (dT) cellulose indicates that it does not possess 3′ polyadenosine tracts (Auperin et al. 1984b). For both viruses, Northern analyses have confirmed that the GPC mRNA species have a viral-sense sequence and that they also correspond in length to approximately half the size of the S RNA species. The GPC mRNA species of LCM virus consists of some 1600 nucleotides, for Pichinde virus the GPC mRNA species is approximately 1610 nucleotides long. These results indicate, therefore, that the two S RNA gene products are coded on separate subgenomic mRNA species and that they must be made from different template RNA species (Fig. 1). Whether the complete arenavirus S RNA species can be translated directly by cellular ribosomes into GPC protein is not known. The existence of a subgenomic GPC mRNA species does not rule out this possibility, although the character of the 5′ end sequence of the viral RNA, or its association with the nucleoprotein, might preclude it.

Although only incomplete sequence data have been reported for the L RNA species of LCM virus (Romanowski and Bishop 1985), the data indicate that the viral L protein is coded in the viral-complementary sequence. So far only one L mRNA species has been identified by Northern analyses (unpublished data). It appears to be similar in size to the viral S RNA. Whether the L RNA has a simple negative-strand strategy, or whether it exhibits an ambisense arrangement, must await complete L RNA sequence data and further molecular analyses.

Comparison of the sequences of the N proteins of Pichinde and LCM viruses indicates that in a best-fit alignment, some 51% of their amino acids are homologous in type and relative position. This homology is a minimal estimate, since it does not take into consideration amino acid differences in the two proteins that represent substitutions of residues of similar charge or character. Bearing this in mind, the data thus indicate that the N proteins of the two viruses probably have similar three-dimensional structures. The GPC primary gene products of Pichinde and LCM viruses exhibit 39% sequence homology, somewhat lower than that observed for the N proteins.

In classical serological procedures such as the complement fixation test, which primarily assesses antigenic epitopes on the viral nucleocapsid protein, different arenaviruses have been shown to be related to each other to different extents (Casals et al. 1975). However, by these procedures LCM virus is reported to exhibit little relationship to members of the Tacaribe complex, although in such tests LCM virus has been shown to be distantly related to Lassa virus (Casals et al. 1975). A report of the DNA sequence of the 3′ half of Lassa virus S RNA (Clegg and Oram 1985) indicates that Lassa viral N protein is related by primary sequence to both LCM and Pichinde viral N proteins. Some 62% of the N protein amino acids of Lassa and LCM viruses and some 50% of the N protein residues of Lassa and Pichinde viruses are identical in type and relative position (Clegg and Oram 1985). This result, as well as the Pichinde – LCM viral N protein relationships, was not expected from the serological experiments or from monoclonal antibody studies which have indicated that few N protein monoclonals of those that have been characterized are cross-reactive between these viruses (Buchmeier et al. 1980, 1981). The fact, though, that some are cross-reactive is in agreement with the reported sequence relationships. The observation that most of the monoclonals that have been analyzed are not cross-reactive may suggest that the conserved regions of the viral N proteins may not be as antigenic as the non-conserved regions. The same may well be true for the viral glycoproteins.

## The Implications of an Ambisense S RNA Strategy

The presence of RNA polymerase activities in arenavirus preparations (Leung et al. 1979) suggests that after attachment, penetration and uncoating in permissive cells, arenaviruses synthesize viral-complementary N and (presumably) L mRNA species (primary transcription). It is assumed that these are translated to give the N and L proteins before RNA replication takes place (Fig. 1). The ambisense strategy of the arenavirus S RNA species precludes the synthesis of the subgenomic S coded GPC mRNA species until after viral RNA replication has been initiated. This is in complete contrast to the strategy employed by members of the negative-stranded rhabdovirus, paramyxovirus and orthomyxovirus groups. For those viruses that have a single species of genomic RNA (i.e., rhabdoviruses and paramyxoviruses) all the viral-complementary, subgenomic, mRNA species are made in a coordinated manner by transcription of the infecting viral RNA species (primary transcription) and, in larger quantities, after viral RNA replication has been initiated (secondary transcrip-

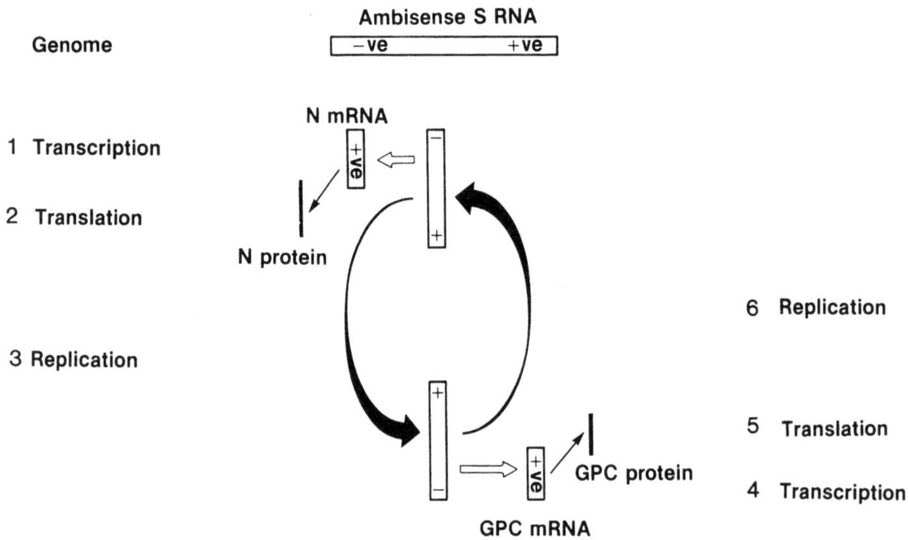


**Fig. 1.** Coding, transcription and replication strategies of arenavirus S RNA species. Whether the viral S RNA also functions as an mRNA is not known

tion) due to the availability of new RNA templates. For the segmented genome orthomyxoviruses all of the viral mRNA species (including those that are derived by splicing) have viral-complementary sequences. Although the selected syntheses of certain influenza mRNA species might be possible, no evidence for such discrimination has been obtained.

The data obtained by analyses of the types and amounts of intracellular S-coded RNA species induced by arenaviruses indicate that the subgenomic N mRNA is made before the subgenomic GPC mRNA species. At later stages of an infection when both S-coded mRNA species are present, the subgenomic N mRNA species predominate (Auperin et al. 1984b; unpublished data). These observations support the hypothesis that the arenavirus N mRNA synthesis is regulated independently of its GPC mRNA synthesis. The obvious advantage of this strategy is that N and (presumably) L proteins are made at the beginning of the infection when they are needed for viral RNA replication. Likewise, GPC mRNA and the viral glycoproteins are not made until they are needed, i.e., after the initiation of RNA replication. Presumably the continued synthesis of N mRNA from the viral RNA template must compete at some stage with viral-complementary, replicative intermediate, RNA synthesis. Nothing is known about the enzymology concerning the process of arenavirus RNA replication or its relationship to the mRNA transcription process. Conceivably different forms of the same gene product (the L protein?) may perform different functions (transcription and replication). The simplest hypothesis, however, is that mRNA transcription and RNA replication are catalyzed by the same enzyme, an enzyme that is influenced in terms of its activity by the availability of newly translated N protein. Notwithstanding these points, it is evident that N mRNA transcription and RNA replication must compete for the same template (similarly L mRNA transcription and L RNA replication). Likewise, the synthesis of GPC mRNA must compete with S viral RNA replication. However, the observation that much less GPC mRNA than S viral RNA species is observed in infected cell extracts, and more N mRNA species is present than S viral-complementary, replicative intermediate RNA, indicates that the determinants of RNA synthesis are complex and not simply regulated just by the form of the enzyme and the availability of N protein.

## The Intergenic Region of the Ambisense S RNA Species

The intergenic regions of Pichinde and LCM viral S RNA species have a unique feature, that of an inverted complementary sequence that may be arranged into a hairpin configuration. For Pichinde virus the intergenic region between N and GPC is 86 nucleotides in length, 36 of which can be arranged into a hairpin configuration (Fig. 2; Auperin et al. 1984 a). For LCM virus the intergenic region is shorter at 64 nucleotides, of which 42 can be arranged into a hairpin structure (Fig. 2; Romanowski and Bishop 1985). The corresponding sequence for Lassa virus has not yet been reported, although it is apparently similar (D. A. Auperin, personal communication). Although no information has yet been published about the mRNA transcription initiation and termination processes for either the S or L RNA species of arenaviruses, in unpublished experiments using oligonucleotides representing viral and viral-complementary sequences of the intergenic region of Pichinde virus S RNA, we have found that transcription of both N and GPC mRNA species terminates near the top of the intergenic hairpin (M. Galinski and D. H. L. Bishop, unpublished results; see Fig. 2). Thus viral-sense oligonucleotides representing the 3′ half of the intergenic hairpin anneal to both N and GPC mRNA species, whereas those representing the 5′ half do not. Both types of oligonucleotide anneal to complete viral S RNA species (viral RNA and replicative intermediate S RNA). How transcription termination in the intergenic region occurs is not known.

## The Origins of Viruses with Ambisense RNA

Concerning the question of how arenaviruses with an ambisense coding strategy may have arisen, unless one invokes an origin from a DNA source in which the arrangement of proteins coded on opposite strands of nucleic acid has been maintained, the simplest explanation is that a chimeric RNA was derived at some stage of arenavirus evolution. Such a chimeric RNA could have been formed during the processes of RNA replication and represent a consolidation of genetic information (i. e., a virus with three species of RNA each coding for a single gene product giving rise to an arenavirus with a consolidated genome and, subsequently, only two RNAs by the exclusion of the redundant third RNA species). Such consolidation of genetic information could occur by a viral replicase copying the coding strand of one RNA species and, instead of terminating, continuing RNA synthesis on the non-coding strand of another RNA. This would result in a chimeric RNA molecule composed of two genes coded on opposite strands of the RNA. Together with the subsequent

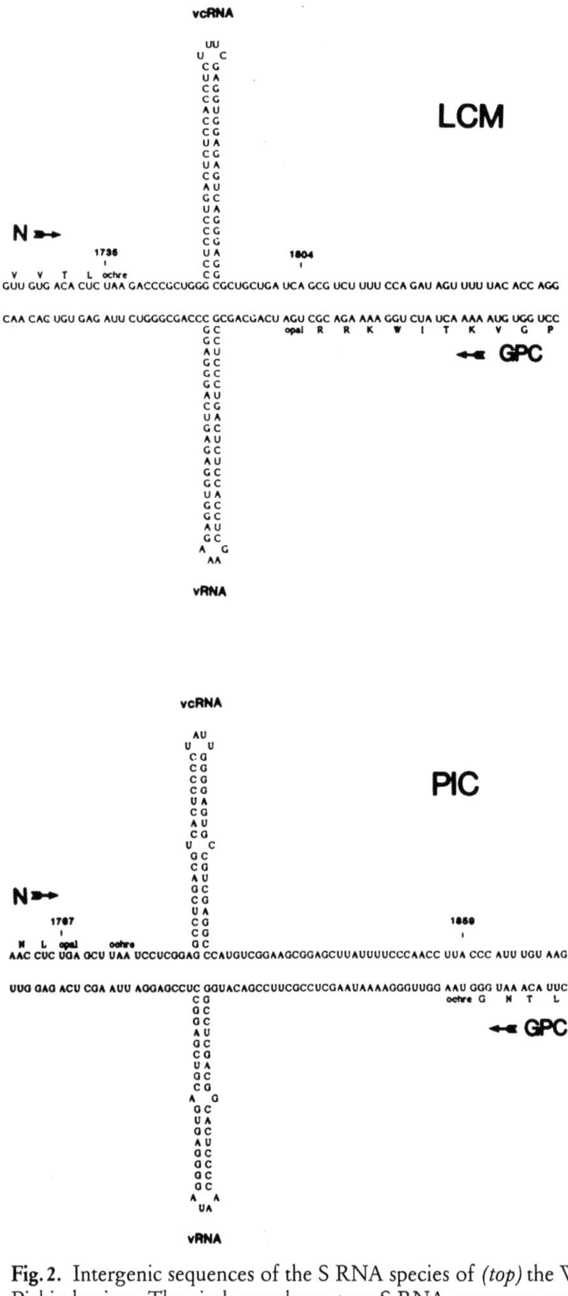


**Fig. 2.** Intergenic sequences of the S RNA species of *(top)* the WE strain of LCM virus and *(bottom)* Pichinde virus. The viral-complementary S RNA sequences are shown above the corresponding viral RNA sequences. The indicated amino acids, translation termination codons and gene products for the two viral S RNA species are shown

loss of the redundant RNA species, a virus would be generated in which all the original genetic information had been conserved. Of course, the reverse situation may have occurred, i.e., the formation of a virus with three RNA species from a virus having two, one of which originally had an ambisense coding arrangement. Whatever the origins, the ambisense coding arrangements that have been observed for arenaviruses (and phleboviruses, Ihara et al. 1984) open yet another dimension to the way in which viruses replicate in cells.

## References

Auperin D, Compans RW, Bishop DHL (1982a) Nucleotide sequence conservation at the 3′ termini of the virion RNA species of New World and Old World arenaviruses. Virology 121: 200–203

Auperin D, Dimock K, Cash P, Rawls WE, Leung WC, Bishop DHL (1982b) Analyses of the genomes of prototype Pichinde and a virulent derivative of Pichinde Munchique arenaviruses: evidence for sequence conservation at the 3′ termini of their viral RNA species. Virology 116: 363–367

Auperin DD, Galinski M, Bishop DHL (1984a) The sequence of the N protein gene and intergenic region of the S RNA of Pichinde arenaviruses. Virology 134: 208–219

Auperin DD, Romanowski V, Galinski M, Bishop DHL (1984b) Sequencing studies of Pichinde arenavirus S RNA indicate a novel coding strategy, an ambisense viral S RNA. J Virol 52: 897–904

Buchmeier MJ, Lewicki HA, Tomori O, Johnson KM (1980) Monoclonal antibodies to lymphocytic choriomeningitis virus react with pathogenic arenaviruses. Nature 288: 486–487

Buchmeier MJ, Lewicki HA, Tomori O, Oldstone MBA (1981) Monoclonal antibodies to lymphocytic choriomeningitis and Pichinde viruses: generation, characterization and cross-reactivity with other arenaviruses. Virology 113: 73–85

Casals J, Buckley SM, Cedeno R (1975) Antigenic properties of the arenaviruses. Bull WHO 52: 421–427

Clegg JCS, Oram JD (1985) Molecular cloning of Lassa virus RNA: nucleotide sequence and expression of the nucleocapsid protein gene. Virology 144: 363–372

Compans RW, Bishop DHL (1985) Biochemistry of arenaviruses. Current Topics in Microbiology and Immunology, vol 114. Springer, Berlin Heidelberg New York, pp 153–175

Harnish DG, Dimock K, Bishop DHL, Rawls WE (1983) Gene mapping in Pichinde virus: assignment of viral polypeptides to genomic L and S RNAs. J Virol 46: 638–641

Howard CR (1986) Arenaviruses. Perspectives in medical virology, vol 2. Elsevier Science, Amsterdam

Ihara T, Akashi H, Bishop DHL (1984) Novel coding strategy (ambisense genomic RNA) revealed by sequence analyses of Punta Toro phlebovirus S RNA. Virology 136: 293–306

Karabatsos N (1985) International catalogue of arboviruses, 3rd edn. American Society of Tropical Medicine and Hygiene, San Antonio

Kirk WE, Cash P, Peters CJ, Bishop DHL (1980) Formation and characterization of an intertypic lymphocytic choriomeningitis recombinant virus. J Gen Virol 51: 213–218

Leung W-C, Rawls WE (1977) Virion-associated ribosomes are not required for the replication of Pichinde virus. Virology 81: 174–176

Leung WC, Leung MKFL, Rawls WE (1979) Distinctive RNA transcriptase, polyadenylic acid polymerase and polyuridylic acide polymerase activities associated with Pichinde virus. J Virol 30: 98–107

Matthews REF (1982) Classification and nomenclature of viruses. Intervirology 17: 1–200

Oldstone MBA, Ahmed R, Buchmeier MJ, Blount P, Tishon A (1985) Perturbation of differentiated functions during viral infection in vivo. 1. Relationship of lymphocytic choriomeningitis virus and host strains to growth hormone deficiency. Virology 142: 158–174

Pedersen IR (1979) Structural components and replication of arenaviruses. Adv Virus Res 24: 277–330

Rawls WE, Leung W-C (1979) Arenaviruses. Comprehensive virology, vol 14. Plenum, New York, pp 157–192
Rawls WE, Chan MA, Gee SR (1981) Mechanisms of persistence in arenaviruses infections: a brief review. Can J Microbiol 27: 568–574
Riviere Y, Ahmed R, Southern PJ, Buchmeier MJ, Dutko FJ, Oldstone MBA (1985) The S RNA segment of lymphocytic choriomeningitis virus codes for the nucleoprotein and glycoproteins 1 and 2. J Virol 53: 966–968
Romanowski V, Bishop DHL (1983) The formation of arenaviruses that are genetically diploid. Virology 126: 87–95
Romanowski V, Bishop DHL (1985) Conserved sequences and coding of two strains of lymphocytic choriomeningitis virus (WE and ARM) and Pichinde arenavirus. Virus Res 2: 35–51
Romanowski V, Matsuura Y, Bishop DHL (1985) The complete sequence of the S RNA of lymphocytic choriomeningitis virus (WE strain) compared to that of Pichinde virus. Virus Res 3: 101–114
Vezza AC, Bishop DHL (1977) Recombination between temperature-sensitive mutants of the arenavirus Pichinde. J Virol 24: 712–715
Vezza AC, Cash P, Jahrling P, Eddy G, Bishop DHL (1980) Arenavirus recombination: the formation of recombinants between prototype Pichinde and Pichinde Munchique viruses and evidence that arenavirus S RNA codes for N polypeptide. Virology 106: 250–260

# Common Events in Mitogenic and Oncogenic Pathways

ROBERT R. FRIIS, ROLF JAGGI, ROLF MÜLLER, ERNST REICHMANN, BRIAN SALMONS[1], ANDREW ZIEMIECKI, and BERND GRONER

Ludwig Institut für Krebsforschung, Inselspital, 3010 Bern, Switzerland

[1] *Present address:* Medical College of Georgia, Department of Cell and Molecular Biology, Augusta, GA 30912, USA

## Introduction

Since activated oncogenes represent spontaneously occurring mutations in growth regulatory genes, it is attractive to employ them in the study of the physiological function of their progenitors. To study the mitogenic effects of oncogenes, systems for their conditional expression are desirable. Unless the oncogenes can be rendered temporarily inactive, no synchronization of cells in $G_0$ is possible.

Transformation-defective temperature-sensitive *(ts)* mutants of Rous sarcoma virus offered the first system for conditional oncogene function. Such mutants could be employed to study the mitogenic effect of *src* gene expression (Bell et al. 1975). With these mutants it could be shown that the *src* gene product was reversibly temperature sensitive: shift to the permissive temperature in the presence of inhibitors of protein synthesis resulted in a rapid rise in tyrosine-specific protein kinase activity (Ziemiecki and Friis 1980). Recently, a different conditional oncogene expression system has been developed (Huang et al. 1981; Jakobovits et al. 1984; Papkoff and Ringold 1984; Jaggi et al. 1986), taking advantage of the glucocorticoid-responsive promoter to the murine mammary tumor virus (Hynes et al. 1983; Ponta et al. 1985).

Following the work of many investigators on parameters relevant to mitogenic events in cells stimulated with defined growth factors [e.g., platelet-derived growth factor (PDGF) and epidermal growth factor (EGF) (Cooper et al. 1982; Stiles 1983; Matrisian et al. 1985)], a set of genes showing induction in early $G_1$ to S of the cell cycle is known. This paper describes a comparison of several early mitogenic events examined in serum or oncogene-stimulated quiescent cells. Experiments directed to elucidating the mechanism of the oncogenes' mitogenic effect will be described.

## Materials and Methods

### RSV transfected Cells

Rous sarcoma virus (RSV) *ts* LA 90-infected NIH 3T3 cells were obtained from Dr. Joan Brugge (Stony Brook, New York). Assays for cell proliferation and ornithine decarboxylase enzyme activity were as previously used by R.R. Friis, R. Jaggi, R. Mueller, J. Forchhammer, and A. Panet (in preparation).

## LTR-*ras*-transfected Cells

The constructions of the mouse mammary tumor virus long terminal repeat (LTR) activated human Harvey *ras* [LTR H-*ras* (A)] plasmid and the LTR normal, nonactivated human Harvey *ras* [LTR H-*ras* (N)] plasmid have been reported (Jaggi et al. 1986). NIH 3T3 cells were transfected with these constructs and single clones isolated. Cell clones transfected with the LTR *ras* (A) construct but not LTR *ras* (N) transfected clones give colonies in semi-solid medium only in the presence of $10^{-6}$ *M* dexamethasone. Within 1 to 3 hours after treatment, transfectants containing both constructs contained high levels of LTR *ras* mRNA. In the absence of the hormone LTR *ras* mRNA is undetectable. The cellular clones employed in these studies were synchronized by culture in 1.5% fetal calf serum for 36 h prior to induction.

## Mammary Epithelial Cells

A construct containing the v-Harvey-*ras* gene in the pZipNeo [SV(x) murine leukemia viral vector (Cepko et al. 1984)] was obtained from Dr. Shelagh Redmond (LICR, Bern) and was employed for transfer of the *ras* gene into the IM-1 ("immortalized" line 1) of mouse mammary epithelial cells. IM-1 is a line of mammary epithelial cells (cytokeratin positive, vimentin negative in immunofluorescence assays) which was isolated by extended cultivation of collagenase-dissociated mouse mammary tissue in medium containing insulin, EGF, and 1% serum. The line shows lactogenic hormone response by reduced growth in the presence of hydrocortisone, insulin, and prolactin, and by the production of small amounts of casein.

## Protein Analysis

In order to detect proteins the expression of which was increased after growth stimulation by serum, or following transformation after shift of RSV *ts* LA 90-infected cells to the permissive temperature, synchronized cells were prepared at the nonpermissive temperature by culture in 1% fetal calf serum-containing medium for 30 h. The growth stimulus was then applied, and 2.5 or 5.5 h later cultures were labeled in methionine-free Dulbecco's modified Eagle's medium with 300 µCi/ml $^{35}$S-methionine (New England Nuclear) for 30 min at the appropriate temperature. Lysates were prepared by Dounce homogenization on ice using Buffer A (20 m*M* $KH_2PO_4$; 1 m*M* NaF; 0.01 m*M* $CaCl_2$; 0.05 m*M* $Na_3Vo_4$; 1 m*M* $\varepsilon$-aminocaproic acid; 1 m*M* PMSF; 2 m*M* $\beta$-mercaptoethanol; 0.05% Triton X-100; 5% glycerol at pH 7.2) supplemented with 1 m*M* EDTA and 0.20% Triton X-100. Following centrifugation at $100000 \times g$ for 30 min, the supernatant was chromatographed on DEAE cellulose beads (Serva) in Buffer A. The flow-through fractions were further purified on the high-performance hydroxyapatite (HPHT) Biorad column at a pressure of 8 bar and flow rate of 0.15 ml/min at 4 °C using a 30-ml linear elution gradient from 20 to 500 m*M* $KH_2PO_4$ in Buffer A. One hundred fractions were collected. A quantity of 30 µl of each fraction was electrophoresed on 12% SDS polyacrylamide gels (24 cm length), silver stained, dried, and exposed for 5 days to autoradiography.

## Results and Discussion

### DNA Synthesis and ODC Activity as a Function of Serum Stimulation and Oncogene Activation

NIH 3T3 *ts* LA 90 cells can be readily synchronized by culture at the nonpermissive temperature (40 °C) for 30 h in medium containing 1% fetal calf serum. Following stimulation with 10% serum, more than 80% of the cells synthesize DNA within 18 h. Less than 20% of unstimulated cells synthesize DNA in the same period (data not shown). In a DNA synthesis experiment using 2-h pulses of $^{125}$I-UdR, the mitogenic effect of the *src* gene product was investigated at permissive (35 °C) and nonpermissive temperatures and compared to the effect of serum (Fig. 1 A). We observed that:

1. The active oncogene product at the permissive temperature is nearly as mitogenic as serum.
2. Unlike serum, its presence is required continuously almost until the beginning of DNA synthesis in order to achieve a maximal percentage of cells in S phase. If the

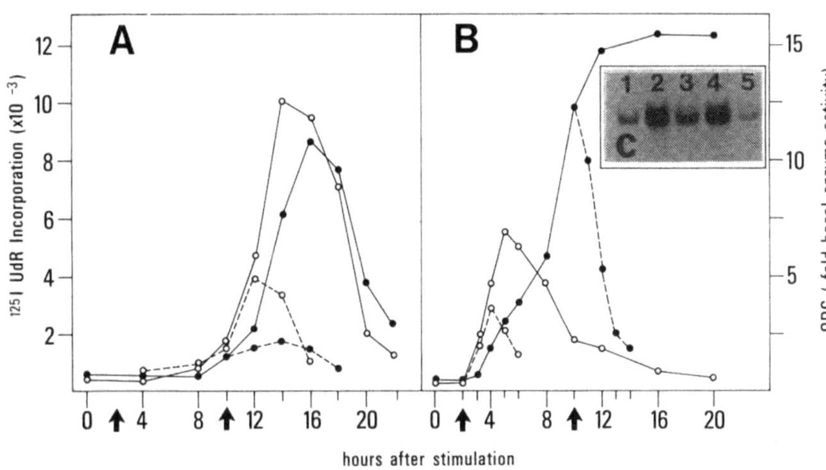


**Fig. 1 A–C. A** The mitogenic response of RSV *ts* LA 90-infected NIH 3T3 cells after stimulation with 10% fetal calf serum at the nonpermissive temperature (40 °C; ○——○) was compared with oncogene stimulation resulting from temperature shift to the permissive temperature (35 °C; ●——●). $^{125}$I-UdR incorporation was measured at different times in 2-h pulses by including 2 μCi/ml in the medium, followed by TCA extraction, and radioisotope determination in a gamma counter. The arrows indicate where the stimuli were removed: serum washed out after 2 h (○----○) or cultures shifted back up to the nonpermissive temperature after 10 h stimulation (●----●). **B** Ornithine decarboxylase *(ODC)* enzyme activity was determined at intervals following the same protocol as for Fig. 1 A. The data are expressed relative to basal activity following normalization to protein concentration in the crude extracts used for assay. The assay was performed as described in Materials and Methods. **C** Total RNA preparations were obtained at several times in an experiment parallel to that of Fig. 1 A, electrophoresed, Northern blotted, and probed by hybridization with ODC-specific SP 6 transcript. Track 1, unstimulated 40 °C; track 2, treated 3 h with 10% fetal calf serum at 40 °C; track 3, 3 h after shift to 35 °C; track 4, 6 h after shift to 35 °C; track 5, 10 h shift to 35 °C, followed by 3 h shift back to 40 °C

oncogene product is inactivated by temperature shift to the nonpermissive temperature as late as 10 h after stimulation, only a small number of cells synthesize DNA. Serum is able to evoke approximately a half-maximal response after a 2-h treatment.

DNA synthesis is a late event in cell proliferation. The early to mid-$G_1$ processes are interesting for the understanding of regulatory mechanisms. Ornithine decarboxylase (ODC) was selected as a enzyme marker for $G_1$ events. The experiment of Fig. 1 A was repeated and samples were taken at different time points. ODC enzyme activity in cellular extracts was assayed by measuring radioactiveity $CO_2$ evolution after reaction with 1-$^{14}$C-ornithine. Both serum and active oncogene product induce ODC activity. The onset of transformation by RSV induces a slowly rising, but ultimately high, constitutive enzyme activity (Fig. 1 B). The same response is observed in unsynchronized, logarithmically growing *ts* LA 90 cells when the temperature is shifted to the permissive temperature (data not shown). Thus, unlike the ODC activity in quiescent cells following serum treatment, the response to the oncogene is not a $G_1$-dependent event, but can occur at any phase of cell cycle.

Total cellular RNA was prepared from cells stimulated as described in Fig. 1. Figure 1 C shows a RNA blot analysis with an ODC-specific probe (McConlogue et al. 1984). In parallel with the enzyme activities, ODC mRNA accumulates to a high level 3 h after serum treatment or oncogene activation. The high level of ODC mRNA is still present in *ts* LA 90 cells after 15 h at the permissive temperature. This illustrates the constitutive nature of the induction, while the "upshift" experiment showing mRNA levels in cells 3 h after return to the nonpermissive temperature demonstrates the dependence of ODC induction on active *src* gene product.

## Cellular Gene Expression is Influenced by *ras* Oncogene Activation

RNA blot analyses were also carried out on RNA samples from NIH 3T3 cells transfected with the LTR-*ras* (A) and LTR-*ras* (N) constructs. Owing to the promoter in the mouse mammary tumor virus LTR, transcription of these constructs is conditionally responsive to glucocorticoid hormones. No *ras* mRNA can be detected in the absence of hormone. Accumulation to a high level is observed within 3 h after hormone addition to the culture medium (data not shown). The cells can be synchronized in the absence of dexamethasone in 1.5% fetal calf serum for 36 h, and a mitogenic effect can be documented resulting in DNA synthesis 12–24 h after dexamethasone stimulation. The *ras* (N) expression does not result in morphological transformation of transfected NIH 3T3 cells, colony formation in agar suspension culture, or tumorigenicity in injected nude mice, but it is as mitogenic as the *ras* (A) (data not shown).

Figure 2 represents mRNA levels in synchronized and in logarithmically growing cells. Cytoplasmic RNA was prepared at five different times after hormonal or serum stimulation. RNA blotting analysis was carried out and hybridized to c-*myc*, ODC, actin, or histone mRNA-specific labeled probes. Hybridization with an H4 histone-specific gene probe shows an increased level of histone mRNA 25 h after hormone addition to synchronized NIH 3T3 cells containing either LTR-*ras* (A) or

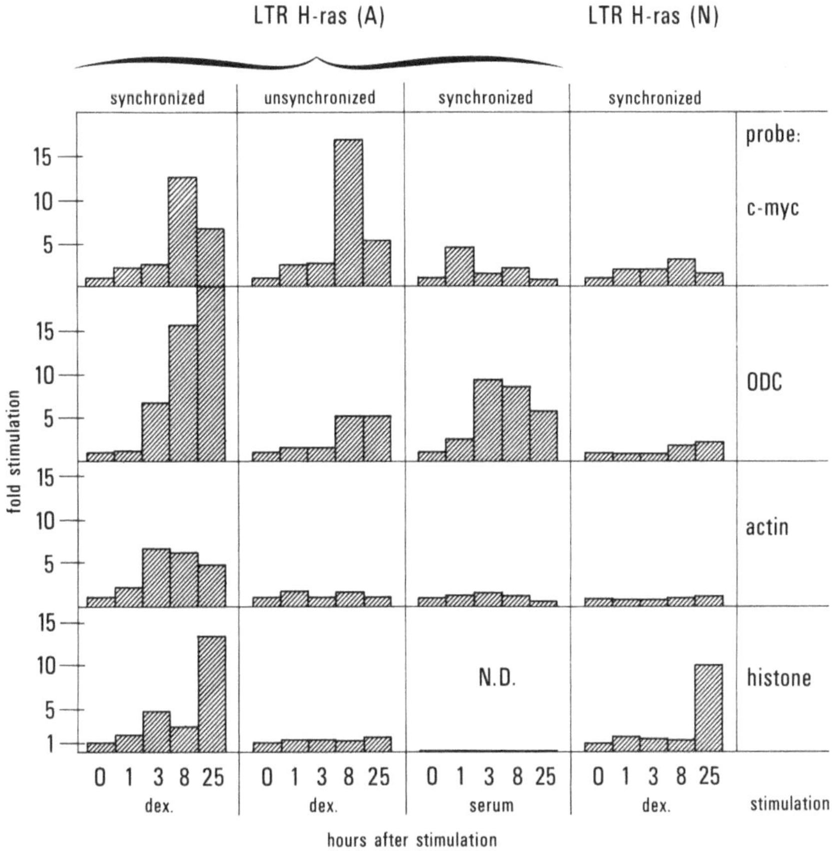


**Fig. 2.** Densitometric profiles from a series of Northern blots performed on cytoplasmic RNA from H-*ras* (A) and H-*ras* (N)-transfected NIH 3T3 cells. The times chosen for RNA preparation were 0, 1, 3, 8, and 25 h after stimulation of the cells with $10^{-6}$ *M* dexamethasone, and the probes were ornithine decarboxylase *(ODC)*-specific, c-*myc*-specific, actin-specific, or H4 histone-specific. The unsynchronized cells received a medium change at intervals of 48 h, while the synchronized cultures were prepared as described in Materials and Methods. *N. D.*, not determined

LTR-*ras* (N), but not in untransfected cells. Logarithmically growing cells express histone mRNA at a constant level in the presence and absence of hormone. The *ras* (A) induction following dexamethasone treatment causes an elevation of c-*myc* and ODC mRNA levels in both synchronized and logarithmically growing cells. Serum stimulation of synchronized cells also causes an increase in c-*myc* and ODC mRNA, but to a lesser degree. c-*myc* mRNA accumulates earlier in serum-stimulated cells. Nonmuscle actin mRNA expression is frequently used as a nonaffected control in gene expression experiments. However, *ras* (A) stimulation of synchronized cells did show an elevated expression of actin mRNA. In contrast to the oncogene, *ras* (N) expression evoked only small responses in c-*myc*, ODC, or actin mRNA levels.

The data of Figs. 1 and 2 indicate that mitogenic stimuli elicited by oncogenes are comparable, but not identical, to those of serum. The responses to oncogenes are more pronounced, i.e., the oncogenes result in a quantitative deregulation of cellular gene expression.

Furthermore, unlike serum, oncogenes evoke their responses in unsynchronized as well as synchronized cells, i.e., they may be able to override the cell cycle regulation of certain gene.

## Effect of *ras* (A) on Differentiation

The mitogenic effect of v-Harvey-*ras* can dominate a differentiation response of mouse mammary epithelial cells to lactogenic hormones, as illustrated in Table 1. The IM-1 line of immortalized mouse mammary epithelial cells undergoes differentiation following lactogenic hormone treatment. This differentiation is manifested by a reduced rate of DNA synthesis, cytoplasmic vacuolization, and production of small amounts of casein.

The *ras* oncogene causes a block of differentiation. The balance between proliferation and differentiation in these lactogenic hormone-responsive cells is disturbed. Constitutive expression of regulatory proteins induced by the oncogene product may prevent entry into a $G_0$ state.

## A Sample of *src*-induced Proteins

Differential screening of cDNA populations has been employed with success to detect unknown genes specifically expressed following stimulation of cells in the $G_0$ phase of the cell cycle to enter the $G_1$ phase (Cochran et al. 1983; Lau and Nathans 1985; Matrisian et al. 1985). Matrisian et al. (1985) were able to functionally identify five genes, namely $\beta$-actin and four glycolytic enzymes, enolase, lactic dehydrogenase, glyceraldehyde phosphate dehydrogenase, and phosphoglycerate mu-

**Table 1.** The influence of v-Harvey-*ras* expression on the lactogenic hormone responsiveness of mouse mammary epithelial cells in culture

| | IM-1 normal | IM-1$^{ras}$ clone 8 | IM-1$^{ras}$ clone 2 |
|---|---|---|---|
| Expression of p21[a] | – | ++ | +++ |
| $^{125}$I-UdR incorporation – HIP[b] (cpm/μg DNA) | 7.1 | 7.7 | 9.3 |
| $^{125}$I-UdR incorporation + HIP (cpm/μg DNA) | 2.8 | 7.3 | 16 |

[a] p21$^{ras}$ protein was measured in protein blots from crude membrane preparations of the different cells, normalized to total protein, electrophoresed on 17% SDS polyacrylamide gels, transferred to nitrocellulose, reacted with p21$^{H\text{-}ras}$-specific rabbit antibody, and detected with $^{125}$I Protein A (New England Nuclear)

[b] TCA-insoluble $^{125}$I-UdR incorporated into the different cells, normalized to DNA content assayed according to Burton et al. (1956), was determined for cells grown without lactogenic hormones, or in the presence of hydrocortisone (5 μg/ml), insulin (10 μg/ml), and prolactin (5 μg/ml) (HIP) in Dulbecco's modified Eagle's medium with 5% fetal calf serum for 4 days prior to assay

tase. Of the more than 60 genes whose expression was detected in early $G_1$, only about a dozen encode biochemically known functions.

We used an approach which might yield information about cell cycle specific protein expression more economically than recombinant DNA cloning. Proteins were pulse-labeled with $^{35}$S-methionine during different intervals in early $G_1$. The proteins were separated by chromatographic and electrophoretic procedures. Figure 3 shows the results of an experiment in which synchronized cells were serum-stimulated and labeled between 2.5 and 3 h after serum addition. Cells stimulated with *src* were labeled between 2.5 and 3 h for one sample, and between 4.5 and 5 h for another. The proteins shown represent a small sample of the total number resolved. The selection is based on solubility in 0.25% Triton X-100 and purification using DEAE chromatography. The DEAE flow-through proteins were further chromatographed using an HPHT Biorad column. Figure 3 shows an autoradiogram of SDS polyacrylamide gel-separated proteins eluted from this column between 50 and 175 m*M* phosphate, pH 7.2.

Three hours after serum treatment (Fig. 3 D), protein bands are evident which are significantly increased in amount relative to unstimulated controls or *src*-stimulated samples. A different set of proteins is responsive to *src* stimulation for 6 h (Fig. 3 B). Three hours stimulation with *src* produces little effect on the cell protein pattern (Fig. 3 A) compared to the unstimulated control. Normalization of the amounts of radioactivity creates a problem in these experiments, since serum treatment stimulates protein synthesis much more than the oncogene, and a correction for total radioactive incorporation mainly reflects rates of synthesis of such major proteins as ribosomal and cytoskeletal components.

The experiment shown in Fig. 3 could be extended to obtain more detailed data. Except for the small aliquot of each HPHT fraction electrophoresed, the samples remain undenatured, so that further purification, for example by affinity chromatography, is possible. Furthermore, since the initial chromatographic steps can be performed with up to 50 mg protein, it is practicable to obtain enough antigen for immunization even from very minor protein species.

## Targets for Tyrosine Phosphorylation by *src* Gene Product

Tyrosine phosphorylation of proteins is associated with oncogenic transformation by RSV (Hunter and Sefton 1980) and many other oncogenes. Phosphotyrosine-containing proteins in RSV-transformed cells are potential targets for transformation-related protein kinases. Table 2 presents a partial list of known target proteins.

Recently, Kamps et al. (1986) have demonstrated a requirement for membrane localization of the *src* gene product for induction of transformation. These authors prepared site-specific mutants in which the amino-terminal glycine of the *src* gene product was substituted. This glycine is essential for acylation by amide linkage to myristic acid. No membrane association for the mutated gene product was observed and the transforming capability was lost. The tyrosine phosphorylation of the candidate targets listed above, however, was unchanged. Hence, the phosphorylation of these proteins is not sufficient, and perhaps not even necessary, for transformation.

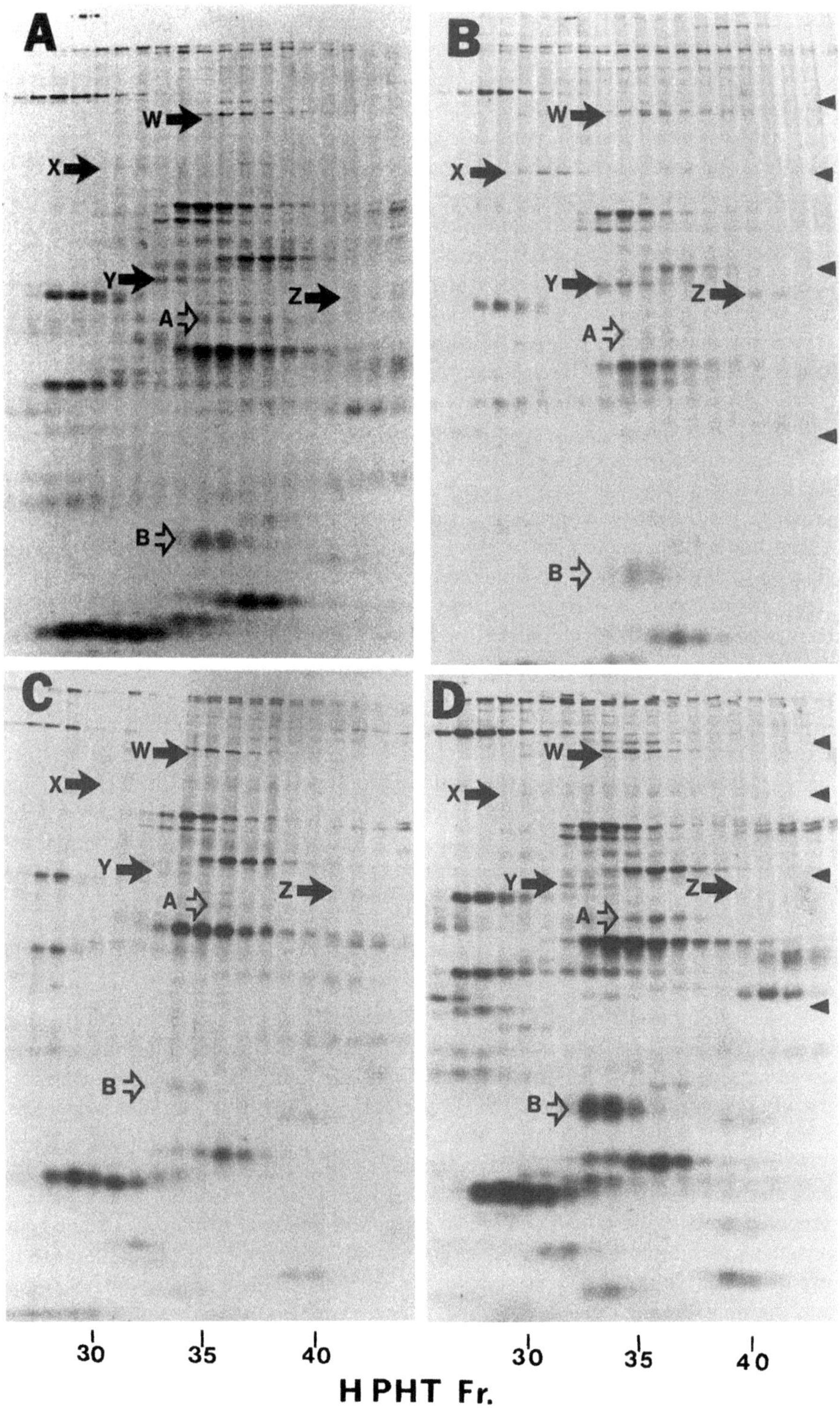
A
B
C
D
W
X
Y
Z
A
B
30
35
40
30
35
40
HPHT Fr.

**Table 2.** Candidate targets for the *src* gene product tyrosine-specific protein kinase

| | Molecular weight | Function | Reference |
|---|---|---|---|
| 36K protein | 36000 | Lipocortin | Radke and Martin (1979) |
| Enolase | 50000 | Glycolytic enzyme | Cooper et al. (1983), Eigenbrodt et al. (1983) |
| Lactic dehydrogenase | 35000 | Glycolytic enzyme | Cooper et al. (1983) |
| Phosphoglycerate mutase | 28000 | Glycolytic enzyme | Cooper et al. (1983) |
| Pyruvate kinase | 60000 | Glycolytic enzyme | E. Eigenbrodt, P. Presek, R. R. Friis, and H. Bauer (unpublished observations) |
| Vinculin | 130000 | Cytoskeletal attachment factor | Sefton and Hunter (1981) |

These experiments compel a new beginning in attempts to understand transformation by RSV and other tyrosine-specific protein kinase oncogenes.

Two considerations might be valuable in this context. First, an attempt must be made to detect very minor targets for tyrosine phosphorylation. Secondly, these targets might be expected to be found among membrane-associated proteins. The methods for the detection of phosphotyrosine-containing proteins are limited by the sensitivity of the labeling procedure and the affinity of phosphotyrosine-specific antibodies. It is therefore desirable to enrich the proteins of interest. Membrane preparation procedures can be used which are sufficiently gentle to avoid the loss of important, but weakly associated proteins. The intensity of the phosphotyrosine signal can also be improved. This can be achieved by an in vitro kinase reaction on the membrane preparations. Under these conditions phosphotyrosine levels 10–20 times higher than in in vivo orthophosphate labeling protocols can be obtained. Conditions can be further optimized for a particular kinase studied (Table 3). Substrates of the oncogene product can be selectively detected, for example by using $Zn^{++}$ as divalent cation for reactions with the *src* gene product or high ATP concentrations for the *fes* oncogene product.

◁ **Fig. 3 A–D.** RSV *ts* LA 90-infected NIH 3T3 cells were synchronized, and parallel cultures were either stimulated by shift for 3 h to the permissive temperature (**A**) or for 6 h to the permissive temperature (**B**), or remained unstimulated as a control at 40 °C (**C**), or received 10% fetal calf serum for 3 h at 40 °C (**D**). In all cases the radioisotopic label was $^{35}$S-methionine for 30 min prior to lysis. Lysates were chromatographed on DEAE cellulose, and the flow-through region was further separated on the Biorad HPHT column. The autoradiograms show the profiles of proteins eluting between 50 and 175 m*M* in the linear phosphate gradient. The proteins designated *W*, *X*, *Y*, and *Z* appear to be disproportionately labeled after shift to the permissive temperature, while the proteins *A* and *B* exhibit a serum-specific expression. Molecular weight markers are shown on the sides of **B** and **D**: 94, 68, 43, and 30 kd

**Table 3.** Parameters distinguishing between different tyrosine-specific protein kinases

| Proteins | Kinase activity[a] with | | |
|---|---|---|---|
| | $Mn^{++}$ (5 m*M*) | $Zn^{++}$ (5 m*M*) | Unlabeled ATP (10 μ*M* to 1 m*M*) |
| c-*src*[b] | + | + | Competition[c] |
| v-*src* | + | + | Competition |
| v-*fgr* | + | − | Competition |
| v-*abl* | + | − | Competition |
| v-*fes* | + | − | Cooperativity[d] |

[a] An in vitro kinase reaction as previously described (Ziemiecki et al. 1984) was performed using $^{32}$P-ATP in the presence of either $Mn^{++}$ or $Zn^{++}$, and with or without added unlabeled ATP. The kinase reaction was performed on immune-precipitated proteins, and phosphorylation was observed for the two *src* proteins on the heavy chain of IgG, and for all the other oncogene proteins as autophosphorylation of the kinase itself

[b] c-*src* designates the normal, proto-oncogene parent of the activated, retroviral oncogene v-*src;* v-*fgr* is the oncogene present in the Theilen-Pedersen isolate of feline sarcoma virus; v-*abl* the oncogene of Abelson murine leukemia virus; v-*fes* the oncogene of the Snyder-Theilen isolate of feline sarcoma virus

[c] Competition denotes a very large loss of radioactive incorporation into the substrate following addition of unlabeled ATP

[d] Cooperativity denotes a small loss or no loss of radioactive incorporation into the substrate following addition of unlabeled ATP, and implies that the kinase $V_{max}$ is greatly increased when a low-affinity ATP-binding site is saturated

## Perspective

The results presented compare and contrast the mitogenic effects of serum and two functionally different oncogenes, *src* and *ras* (A). Biochemically, both stimuli are mitogenic. Important differences in the pattern of gene expression, especially for ODC, accompany the oncogene stimulus. Both oncogenes induce changes in gene expression in unsynchronized, logarithmically growing cells. It is likely that other cellular proteins are also released from cell cycle regulation. A search for other early $G_1$ proteins using protein purification methods and short pulse labels promises to yield systematic information about regulatory events taking place in this critical period of cell growth.

Eventually it will be necessary to discover the intermediate biochemical events lying between stimulus and the beginning of DNA synthesis. The differences in gene expression between serum-stimulated and oncogene-stimulated cells discussed above reflect a response to transmission of a message from the membrane location of effectors in both serum and oncogene pathways. Receptors for such growth factors as PDGF, EGF and insulin have been shown to possess a tyrosine-specific protein kinase activity (Nishimura et al. 1982; Shia and Pilch 1983; Downward et al. 1984). It remains hypothetical that a tyrosine-phosphorylated substrate is the mediator of the message. Recent experience, however, has shown that a vastly increased sensitivity to minor tyrosine phosphorylations can be developed, perhaps resulting in the identification of the substrate protein relevant for mitogenic signal transmission.

## Summary

Retroviral oncogenes are mutated versions of host genes thought to be responsible for regulation of cell differentiation and proliferation. Progress in understanding the control mechanisms for cell proliferation has been rapid in recent years because quiescent cell systems responding to specific growth factors by synchronous entry into cell cycle have been developed. Temperature-sensitive mutants of the Rous sarcoma virus *src* gene and glucocorticoid-responsive constructs of the mouse mammary tumor virus long terminal repeat human H-*ras* gene provide two different conditional expression systems for oncogenes. Cells can be synchronized in $G_0$ before activation of oncogene function. Mitogenic events and transformation-specific events can be differentiated. Onset of transformation can be induced with the activation of the *src* or *ras* oncogenes. The biochemical consequences can be compared to those observed after serum stimulation of quiescent cells. Similarities and differences are observed: both oncogenes and serum induce cell DNA synthesis and H4 histone gene expression with similar kinetics. The c-*myc* and ornithine decarboxylase mRNA pools are elevated. Induction of c-*myc* and ornithine decarboxylase expression occurs later and is stronger following transformation than following serum stimulation.

Both *src* and *ras* induce cell cycle-independent and enhanced, constitutive expression of the ornithine decarboxylase gene at the level of mRNA concentration and enzyme activity. Furthermore, *ras* expression provides a constitutive signal for proliferation in lactogenic hormone-responsive mouse mammary epithelial cells. Treatment of mouse mammary epithelial cells with lactogenic hormones inhibits proliferation and cells differentiate as indicated by casein synthesis. Expression of *ras* in these cells abolishes differentiation and stimulates proliferation.

The modified pattern of gene expression of transformed cells can be confirmed by examining cellular proteins by chromatographic procedures. A number of proteins induced as a consequence of *src* function were found. These proteins are not seen following serum stimulation. Serum-induced proteins not seen following transformation were also observed.

The tyrosine-specific protein kinase activity shared by the *src* gene product and several peptide hormone receptors offers an important approach to learning about the initial steps of growth induction, by identification of relevant substrate proteins. This unusual kinase activity may be responsible for the "second signal" that transmits the stimulus to the cell nucleus, which brings about the changes in gene expression discussed above. New approaches are needed to find relevant substrates which become phosphorylated at the plasma membrane.

*Acknowledgements.* We thank Dr. Raymond Sweet, Smith Kline and French Laboratories, for making $p21^{ras}$-specific antibodies available to us, and Dr. Phillip Coffino, University of California, San Francisco, for giving us a cDNA clone of ornithine decarboxylase.

## References

Bell JG, Wyke JA, Macpherson IA (1975) Transformation by a temperature sensitive mutant of Rous sarcoma virus in the absence of serum. J Gen Virol 27: 127–134

Burton K (1956) A study of the conditions and mechanism of the diphenylamine reaction for the colorimetric estimation of deoxyribonucleic acid. Biochem J 62: 315–323

Cepko CL, Roberts BE, Mulligan RC (1984) Construction and applications of a highly transmissible murine retrovirus shuttle vector. Cell 37: 1053–1062

Cochran BH, Reffel AC, Stiles CD (1983) Molecular cloning of gene sequences regulated by platelet-derived growth factor. Cell 33: 939–947

Cooper JA, Bowen-Pope DF, Raines E, Ross R, Hunter T (1982) Similar effects of platelet-derived growth factor and epidermal growth factor on the phosphorylation of tyrosine in cellular proteins. Cell 31: 263–273

Cooper JA, Reiss NA, Schwartz RJ, Hunter T (1983) Three glycolytic enzymes are phosphorylated at tyrosine in cells transformed by Rous sarcoma virus. Nature 302: 218–223

Downward J, Yarden Y, Mayes E, Scrace G, Totty N, Stockwell P, Ullrich A, Schlessinger J, Waterfield MD (1984) Close similarity of epidermal growth factor receptor and v-*erb* B oncogene protein sequences. Nature 307: 521–527

Eigenbrodt E, Fister P, Ruebsamen H, Friis RR (1983) Influence of transformation by Rous sarcoma virus on the amount, phosphorylation and enzyme kinetic properties of enolase. EMBO J 2: 1565–1570

Huang AL, Ostrowski AC, Berard D, Hager GC (1981) Glucocorticoid regulation of the H-MuSV $p^{21}$ gene conferred by sequences from mouse mammary tumor virus. Cell 27: 245–255

Hunter T, Sefton BM (1980) The transforming gene product of Rous sarcoma virus phosphorylates tyrosine. Proc Natl Acad Sci USA 77: 1311–1315

Hynes N, van Ooyen AJJ, Kennedy N, Herrlich P, Ponta H, Groner B (1983) Subfragments of the large terminal repeat cause glucocorticoid-responsive expression of mouse mammary tumor virus and of an adjacent gene. Proc Natl Acad Sci USA 80: 3637–3641

Jaggi R, Salmons B, Muellener D, Groner B (1986) The v-*mos* and H-*ras* oncogene expression represses glucocorticoid hormone dependent transcription from the mouse mammary tumor virus LTR. EMBO J 5: 2609–2616

Jakobovits EB, Majors JE, Varmus HE (1984) Hormonal regulation of the Rous sarcoma virus *src* gene via a heterologous promoter defines threshold dose for cellular transformation. Cell 38: 565–575

Kamps MP, Buss JE, Sefton BM (1986) Rous sarcoma virus transforming protein lacking myristic acid phosphorylates known polypeptide substrates without inducing transformation. Cell 45: 105–117

Lau LF, Nathans D (1985) Identification of a set of genes expressed during $G_0/G_1$ transition of cultured mouse cells. EMBO J 4: 3145–3151

Matrisian LM, Rautmann G, Magun BE, Breathnach R (1985) Epidermal growth factor or serum stimulation of rat fibroblasts induces an elevation in mRNA levels for lactate dehydrogenase and other glycolytic enzymes. Nucleic Acids Res 13: 711–726

McConlogue L, Gupta M, Wu L, Coffino P (1984) Molecular cloning and expression of the mouse ornithine decarboxylase gene. Proc Natl Acad Sci USA 81: 540–544

Nishimura J, Huang JS, Deuel TF (1982) Platelet-derived growth factor stimulates tyrosine-specific protein kinase activity in Swiss mouse 3T3 cell membranes. Proc Natl Acad Sci USA 79: 4303–4307

Papkoff J, Ringold GM (1984) Use of the mouse mammary tumor virus long terminal repeat to promote steroid-inducible expression of v-*mos*. J Virol 52: 420–430

Ponta H, Kennedy N, Skroch P, Hynes NE, Groner B (1985) Hormonal response region in the mouse mammary tumor virus long terminal repeat can be dissociated from the proviral promoter and has enhancer properties. Proc Natl Acad Sci USA 82: 1020–1024

Radke K, Martin GS (1979) Transformation by the Rous sarcoma virus: effects of *src* gene expression on the synthesis and phosphorylation of cellular polypeptides. Proc Natl Acad Sci USA 76: 5212–5216

Sefton BM, Hunter T (1981) Vinculin: a cytoskeletal target of the transforming protein of Rous sarcoma virus. Cell 24: 165–174

Shia MA, Pilch PF (1983) The beta subunit of the insulin receptor is an insulin-activated protein kinase. Biochemistry 22: 717–721
Stiles CD (1983) The molecular biology of platelet-derived growth factor. Cell 33: 653–655
Ziemiecki A, Friis RR (1980) Phosphorylation of pp60$^{src}$ and the cycloheximide insensitive activation of the pp60$^{src}$-associated kinase activity of transformation-defective temperature sensitive mutants of Rous sarcoma virus. Virology 106: 391–394
Ziemiecki A, Hennig D, Gardner L, Ferdinand FJ, Friis RR, Bauer H, Pedersen NC, Johnson L, Theilen GH (1984) Biological and biochemical characterization of a new isolate of feline sarcoma virus: Theilen-Pedersen (TP1-FeSV). Virology 138: 324–331

# Molecular Basis for the Pathogenicity of the Friend Spleen Focus-Forming Virus

LINDA WOLFF, SIU-WAH CHUNG, and SANDRA RUSCETTI

Laboratory of Genetics, National Cancer Institute, National Institutes of Health, Bethesda, MD 20892, USA

## Introduction

Murine retroviruses that induce leukemia in mice can be classified into two groups depending upon their potency in adult animals. Viruses in class I (see Table 1) cause leukemia only after long induction periods which can be explained by their indirect mechanisms of action; these often involve rare integrations that activate transcription of cellular proto-oncogenes. The more acute activity of the class II viruses is evidenced not only by their short induction periods for leukemia in vivo but also by their ability to transform hematopoietic cells in culture. The transforming activity of the majority of viruses shown in class II (see Table 1) can be explained by the presence in the genome of a cellularly derived oncogene. However, the transforming activity of the spleen focus-forming virus (SFFV), which is the subject of this review, cannot be explained in this way. The SFFV genome is unique in that it does not contain an oncogene but is composed entirely of retroviral sequences. Our interest in the transforming nature of this virus has led us to dissect the viral genome in order to determine the location of the pathogenic elements.

**Table 1.** Murine leukemia viruses

| | Disease | Contains oncogene |
|---|---|---|
| Class I: slow onset in adults (> 3 months) | | |
| Friend MuLV | Erythroid | – |
| Rauscher MuLV | Erythroid | – |
| Moloney MuLV | Lymphoid | – |
| AKR MuLV | Lymphoid | – |
| Class II: rapid onset in adults (< 2 months) | | |
| Abelson MuLV | Lymphoid | + |
| Harvey MSV | Erythroid | + |
| Kirsten MSV | Erythroid | + |
| Myeloproliferative sarcoma virus | Multiple | + |
| SFFV | Erythroid | – |

MuLV, murine leukemia virus; MSV, murine sarcoma virus; SFFV, spleen focus-forming virus

The Friend SFFV causes an acute erythroblastosis in both newborn and adult mice which is manifested at 9 days post infection by foci on the spleen and at 2–3 weeks by a grossly enlarged liver and spleen (Friend 1957; Axelrad and Steeves 1964; Mirand et al. 1968; reviewed in Ruscetti and Wolff 1984). This phase represents the hyperblastic response of erythroid precursor cells to the virus. If mice live long enough (more than 3 weeks), one can detect in their spleens more autonomous malignant cells that represent a second stage in the transformation process (Wendling et al. 1981; Mager et al. 1981). These cells, unlike the first-stage cells, are monoclonal and probably represent progeny of a single cell that has undergone a secondary genetic event. The "Friend virus complex," which has traditionally been used to induce the disease, is composed of a replication-defective component, SFFV, and a replication-competent component (Fig. 1). The latter is most often Friend murine leukemia virus (F-MuLV), but it can also be other competent helper viruses such as Moloney or Rauscher murine leukemia virus (M-MuLV, R-MuLV).

Initial experiments designed to locate the pathogenic region of SFFV were carried out a few years ago by Linemeyer et al. (1981, 1982). Subgenomic fragments of DNA and genetically engineered deletion mutants were prepared and tested for biological activity. They were tested by cotransfecting the DNA into fibroblasts along with helper virus DNA, recovering virus particles produced by the cells and inoculating them into mice. Results showed that when the DNA that was used for transfection included an intact segment from the 3′ region of the SFFV genome (designated by the line above SFFV in Fig. 1), mice developed erythroleukemia. However, if this region was not complete or contained a deletion, the mice did not develop disease. It was suggested from these studies that the envelope gene, which is a major part of the designated region, is essential for initiation of the leukemogenic process.

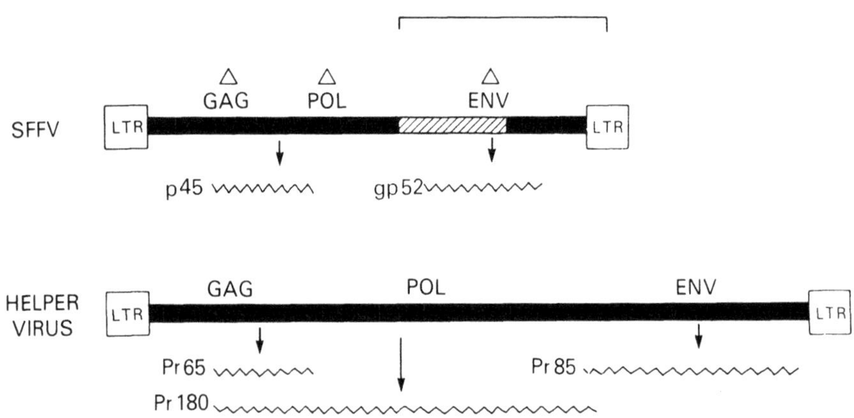


**Fig. 1.** Genome structures of viruses that comprise the Friend virus complex. The SFFV genome and the helper virus genomes are composed of either complete or deleted ($\Delta$) *gag*, *pol*, and *env* genes (see Weiss et al. 1982 for review of retrovirus structures). Although the helper virus produces all of the normal precursor proteins, including $Pr65^{gag}$, $Pr180^{gag\text{-}pol}$ and $gPr85^{env}$, the SFFV produces only two proteins, p45 and gp52, which are defective for replicative function. The *hatching* designates the region which contains endogenously acquired sequences. The *line* above the SFFV genome indicates the pathogenic region as defined by Linemeyer et al. (1981, 1982). *LTR*, long terminal repeat

Despite these results, questions concerning the function of other parts of the Friend virus complex in the leukemogenic process remained unanswered. Herein we describe studies designed to assess the role in disease of the enhancer-containing long terminal repeat (LTR) region of SFFV, to assess the role of the helper virus and to further characterize specific sequences within the evelope gene.

## Results and Discussion

### Role of the Helper Virus in SFFV Disease

The requirement for helper virus in SFFV disease was tested by preparing SFFV stocks that were devoid of replication-competent virus. By using such preparations to infect mice we were attempting to answer the following questions: (1) Do helper-virus gene products cooperate with the envelope gene product of SFFV, gp52, to produce the oncogenic phenotype? (2) Is replication and spread of SFFV required for the development of the progressive fulminating disease? (3) Is the helper virus required for the second, more autonomous stage of transformation?

Helper-free SFFV was prepared by using $\psi$-cells, which contain a mutant M-MuLV genome lacking the sequences required for packaging of its own RNA into viral particles (Mann et al. 1983). Cells transfected with SFFV DNA were shown to produce retroviral particles containing only SFFV RNA, since NIH 3T3 cells infected with culture fluids from these cells express SFFV, but not MuLV, envelope proteins and do not release reverse transcriptase or infectious virus (Wolff and Ruscetti 1985). When helper-free SFFV particles were tested for their biological effects both in vitro, in an erythroid transformation assay, and in vivo, by direct intravenous inoculation of mice, they were found to behave similarly to helper-containing stocks of SFFV. As shown in Table 2, it was necessary to pretreat mice with phenylhydrazine to increase the number of target cells available to the virus in order to develop disease. One hundred percent of the mice injected with helper-free SFFV showed foci of transformed cells on their spleens at 9 or 12 days post inoculation. When mice were examined at 20 and 27 days, it was found that 50% had progressive erythroleukemia and it was assumed that the other 50% had regressed their tumors. Spleens of mice with progressive leukemia were indistinguishable histologically from those of mice given helper virus-containing stocks of SFFV, and they expressed high levels of the SFFV envelope glycoprotein, gp52, but not helper virus-encoded proteins. No infectious virus could be detected in these spleens. These results indicate that (1) expression of helper virus genes in erythroid target cells is not necessary in conjunction with SFFV for the transformation process, and (2) systemic infection which allows continued introduction of SFFV into new target cells is not required for fulminating disease.

Previous studies have shown that erythroid cells transformed by SFFV, in the presence of helper virus, can undergo multiple stages of malignancy giving rise to immortalized erythroid cells that are able to grow indefinitely either in vivo or in vitro. In order to determine whether helper virus is required for conversion of first-stage cells to cells having a more autonomous or malignant phenotype, we tested the

**Table 2.** Effects of helper-free SFFV in vivo

| Pretreatment | Virus[a] | Day p.i. | No. positive/ no tested[b, c] | Average spleen wt in grams (range) | Splenic foci | Early disease (%)[b] | Progressive disease (%)[c] |
|---|---|---|---|---|---|---|---|
| Phenyl-hydrazine | None | 12 | 0/5 | 0.14 (0.13–0.16) | – | 0 | 0 |
| | | 20 | 0/3 | 0.12 (0.11–0.13) | – | 0 | 0 |
| | ψ-2 | 12 | 0/5 | 0.18 (0.13–0.35) | – | 0 | 0 |
| | | 20 | 0/3 | 0.12 (0.10–0.14) | | | |
| | $SFFV_P$ (ψ-2) | 9 | 3/3 | ND | + | 100 | |
| | | 12 | 3/3 | 0.35 (0.15–0.60) | + | | |
| | | 20 | 3/5 | 1.12 (0.14–1.83) | | | 50 |
| | | 27 | 1/3 | 0.73 (0.10–1.94) | | | |
| None | $SFFV_P$ (ψ-2) | 12,20 | 0/13 | 0.10 (0.09–0.10) | – | 0 | 0 |

[a] Mice were infected with supernatant from ψ-2 cells or ψ-2 cells transfected with $SFFV_P$ DNA
[b] Mice were considered positive for early SFFV disease if the spleens had foci at day 9 or day 12
[c] Mice were considered positive for progressive hyperblastic disease if the spleen weight was greater than 0.4 g on day 20 or day 27

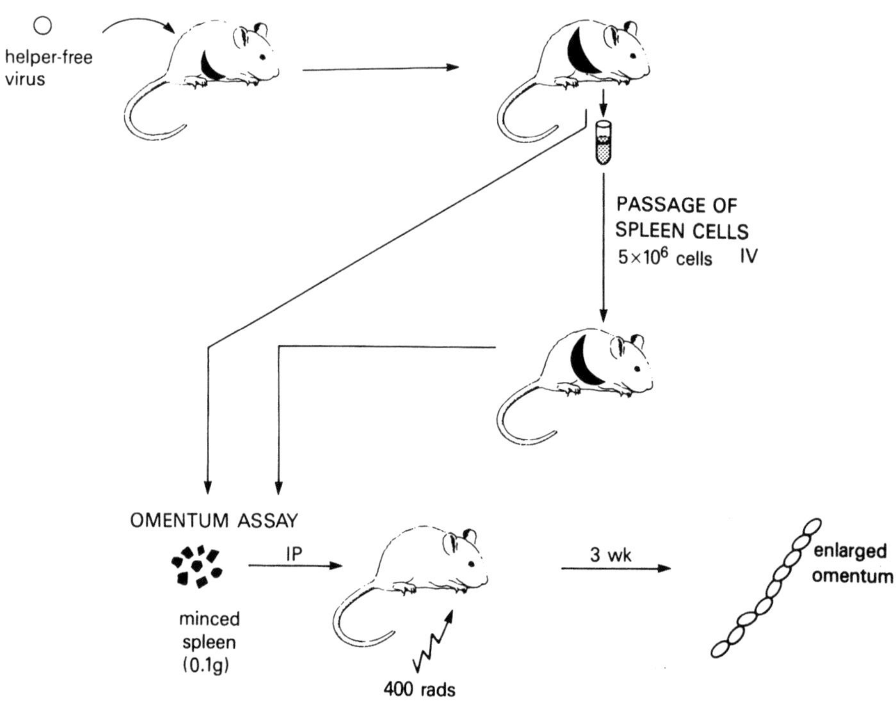


**Fig. 2.** In vivo demonstration of malignant erythroid cells

tumorigenicity of in vivo derived erythroid cell populations transformed by helper-free SFFV (Wolff et al. 1986). The results indicate that spleens enlarged as a result of helper-free SFFV transformation will give rise after several weeks to cells which can be transplanted serially into the omentum of syngeneic sublethally irradiated recipients as outlined in Fig. 2. The autonomous nature of the cells derived from such a spleen could also be demonstrated by the ability of these cells to grow indefinitely as erythroleukemic cell lines. These lines, like their helper virus-containing counterparts, grow as subcutaneous tumors in syngeneic mice, are free of infectious virus, express SFFV but not F-MuLV viral proteins, express the transformation-related protein p53 and can be induced to differentiate into hemoglobin-containing cells using dimethylsulfoxide and hexamethylenebisacetamide but not erythropoietin. It is, therefore, clear from these studies that replication-competent helper virus is not required for the development of the second stage of SFFV-induced transformation. Since SFFV alone can induce both first- and second-stage erythroid disease and the only virus component present in cells from the different stages is SFFV, one must conclude that SFFV is either directly or indirectly responsible for induction of both phases.

## Role of the Long Terminal Repeat in SFFV Disease

Further experiments were designed to test whether the SFFV LTR was specifically required for pathogenicity or could be substituted by other LTR sequences from other retroviruses. In particular, we wanted to examine the function of the LTR in determining erythroid tissue specificity, since it had previously been shown by Chatis et al. (1983, 1984) and others (Ishimoto et al. 1985; Vogt et al. 1985) that the LTRs of F-MuLV and Friend mink cell focus-inducing (F-MCF) viruses contained sequences responsible for erythroid leukemia, while the LTR of M-MuLV is responsible for lymphoid leukemia. Another question addressed in the LTR experiments was the necessity of having one copy of the direct repeat in the enhancer region, since unlike most retroviruses, which have two copies, SFFVs have only one copy of the direct repeat (Clark and Mak 1982; Bestwick et al. 1984; Wolff et al. 1985).

Recombinant SFFVs in which the 3′ LTR was substituted by other viral LTRs were prepared and are shown in Fig. 3 (Wolff and Ruscetti 1986). LTR fragments used in this experiment were derived from erythroleukemia-inducing viruses, F-MuLV and F-MCF, and from lymphoma-inducing viruses, M-MuLV and M-MCF. Both of the MuLVs have two copies of the direct repeat, whereas the MCF viruses have only one. In vivo testing of the recombinant viruses was carried out by injecting adults with F-MuLV pseudotypes (recombinant viruses plus F-MuLV helper virus) or injecting newborns with amphotropic virus pseudotypes. It was found that all of the recombinant viruses produced erythroleukemia in 85% or more of the mice (Fig. 4). It was clear in this study that the spleens of leukemic mice were infiltrated with erythroblasts and not lymphoblasts, since the spleen cells responded to the erythroid hormone, erythropoietin (Epo), in a $^3$H-thymidine incorporation assay and the spleens from infected mice had higher numbers of CFU-E (2-day erythroid colonies) than normal spleens (Wolff and Ruscetti 1986). From these results we concluded that it was inconsequential whether there are one or two copies of the direct

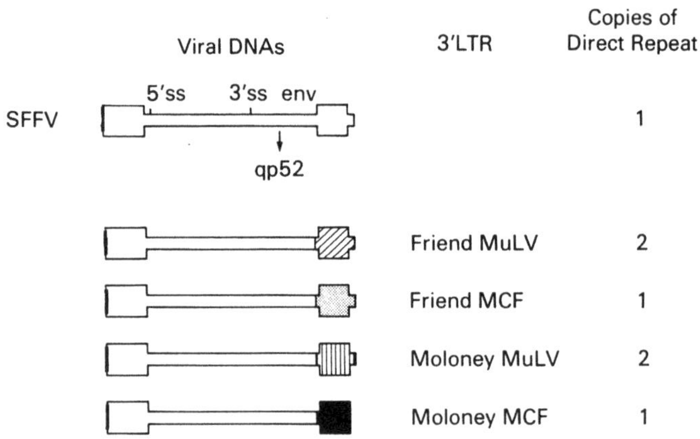


**Fig. 3.** Recombinant SFFVs with new LTRs. LTR fragments from the viruses indicated were ligated onto the 3′ end of the SFFV genome. *ss,* splice site

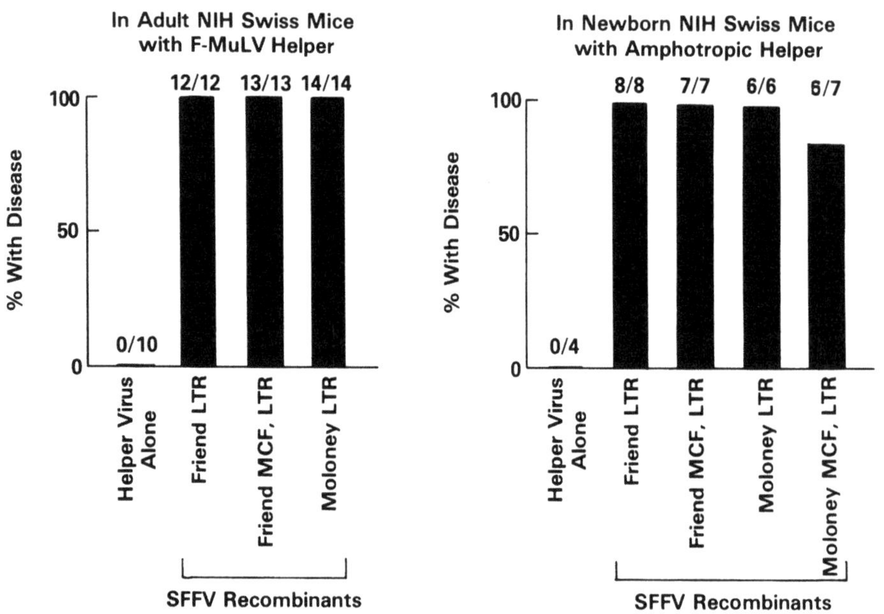


**Fig. 4.** Induction of erythroleukemia by SFFV-LTR recombinants. *Bars* represent the percentage of mice with erythroleukemia. Disease incidence is indicated above the bars as the number of mice that had progressive leukemia per number of mice infected

repeat in the enhancer region of the SFFV LTR and that the LTR of SFFV, unlike those of other murine leukemia viruses, does not specify the tissue tropism of the disease. It appears, then, that the part of the virus exclusively responsible for inducing erythroid cell proliferation is the coding region for the SFFV envelope protein, gp52.

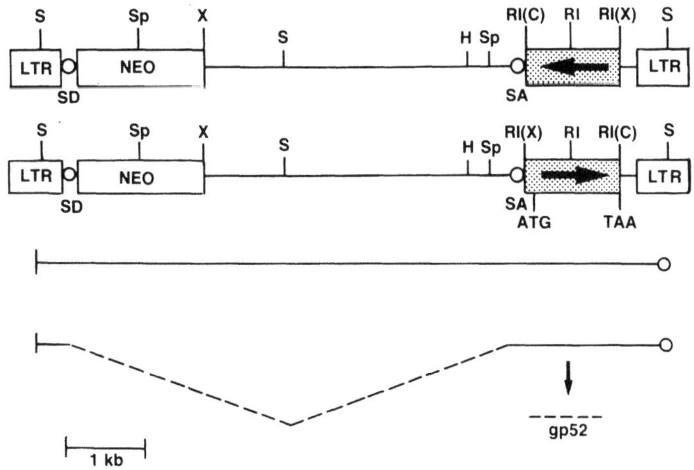


**Fig. 5.** Moloney MuLV vectors containing the coding region of gp52. The *stippled area* represents the gp52 coding region and the *arrow* indicates the transcriptional orientation. The protein gp52 is transcribed from a subgenomic messenger RNA while the product of the neomycin *(NEO)* resistance gene, at the 5′ end, is translated from the full-size genomic message. Restriction endonuclease sites in parentheses were converted to *Eco*RI sites by linkering. *SD* splice donor; *SA*, splice acceptor; *C*, *Cla*I; *H*, *Hin*dIII; *RI*, *Eco*RI; *S*, *Sst*I; *Sp*, *Sph*I; *X*, *Xma*III

## Pathogenicity of a M-MuLV Vector Containing the Coding Region for gp52

To test the hypothesis that the envelope protein of SFFV, gp52, is the sole transforming element of the virus, the gp52 coding region DNA was isolated and placed in an M-MuLV-based vector (Fig. 5) (Wolff and Ruscetti, 1988). It was inserted in either the correct or the reverse transcriptional orientation. The vectors, which were developed by Gilboa et al. (1982), contain the neomycin resistance gene which was used to select cells transfected with the vector DNA and to quantitative virus produced by cells. The vector with the SFFV DNA in the correct orientation, when rescued with M-MuLV, produced erythroleukemia in 49 of 55 mice tested, whereas that with the DNA in the reverse orientation did not induce disease. These results provide the firmest evidence to date that the *env* gene of SFFV is solely responsible for its pathogenicity.

## Molecular Basis for the SFFV$_P$ and SFFV$_A$ Phenotypes

Many previous studies have demonstrated that the primary effect of SFFV on erythroid cells is to alter the requirement that the cells have for the erythroid-specific hormone, erythropoietin (for review see Ruscetti and Wolff 1984), and the studies described above indicate that the viral *env* gene product is responsible for the effect. To further elucidate how the viral *env* gene product may do this, we have been studying variants of the virus, SFFV$_P$ and SFFV$_A$, which differ in their requirements for erythropoietin to divide and differentiate and which produce envelope gene products that differ in their post-translational modification. Cells infected with the polycythemia-inducing strain ($SFFV_P$) but not the anemia-inducing strain ($SFFV_A$) appear to be erythropoietin-independent. In other words, progenitor erythroid cells infected with the former virus proliferate and differentiate maximally in the absence of the hormone, whereas $SFFV_A$-infected cells require the hormone for maximum stimulation and will not become hemoglobinized (differentiated) in its absence. The two viruses differ in the post-translational modification of the product of their *env* genes, because the *env* gene product of $SFFV_P$ is processed to a higher-molecular-weight form by the addition of complex carbohydrates and can be detected on the cell surface, whereas the $SFFV_A$ *env* protein is not detectable on the cell surface (Ruscetti et al. 1981). Sequence analysis of the *env* gene of both strains has shown that the majority of amino acid differences reside in the C-terminal half of the gp52 molecule (Wolff et al. 1985). Many of these differences are in the p15E anchoring portion that transverses the cellular membrane.

$SFFV_A/SFFV_P$ recombinant viruses were prepared with these sequence differences in mind (Fig. 6). Two of these constructs have 3′ envelope and LTR sequences ($SFFV_{AP\text{-}L}$) or only 3′ envelope sequences ($SFFV_{A\text{-}RC}$) derived from the P variant and the remainder of the genome from the A strain. Interestingly, it was found that both of these recombinant viruses possess all of the biological as well as biochemical characteristics of $SFFV_P$ (Wolff and Ruscetti 1985; Ruscetti and Wolff 1985; Chung et al. 1987). Results from erythroid colony assays indicated that cells infected with this virus can differentiate without added Epo, and high proliferation was also measured in the absence of Epo in a $^3$H-thymidine assay. Further studies indicated that the envelope glycoprotein encoded by this recombinant virus was properly glycosylated to the higher-molecular-weight form that eventually appears on the cell surface. All of these are unique features that characterize $SFFV_P$-infected cells. Our da-

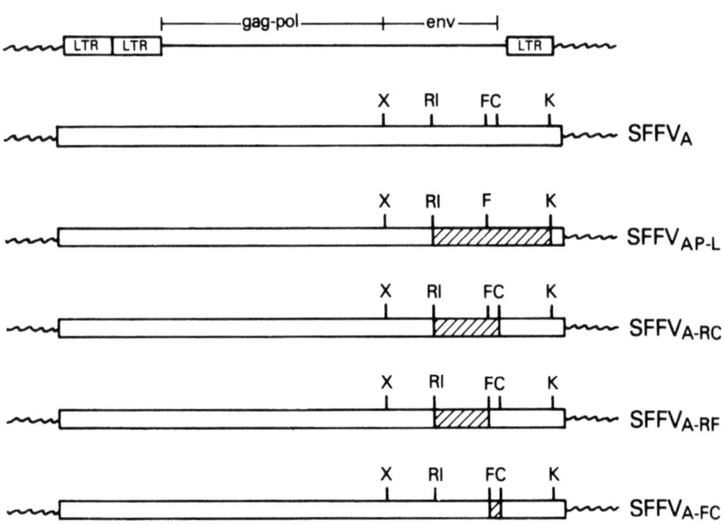


**Fig. 6.** $SFFV_A/SFFV_P$ recombinant viruses. Sequences derived from the $SFFV_P$ strain are designated by a *hatched bar* while those from the A strain are designated by an *open bar*. *C*, *Cla*I; *F*, Fok*I;* *K*, *Kpn*I; *RI*, *Eco*RI; *X*, *Xba*I. For references to construction of $SFFV_A$, $SFFV_{AP\text{-}L}$, and $SFFV_{A\text{-}RC}$ see Wolff and Ruscetti (1985) and Chung et al. (1987)

ta, therefore, provide direct evidence that the 678-bp fragment at the 3′ end of the SFFV$_P$ *env* gene is responsible for the biological and biochemical differences between the two viral strains. We are now further dissecting this region into two subregions using the *Fok*I restriction site. This site divides the 3′ end of the SFFV *env* gene into *Eco*RI-*Fok*I and *Fok*I-*Cla*I fragments, the latter of which contains 120 bp from the extreme 3′ end of the *env* gene which codes for the hydrophobic portion of the protein where the majority of the sequence differences between SFFV$_P$ and SFFV$_A$ reside. Recombinant viruses that contain either the *Eco*RI-*Fok*I or the *Fok*I-*Cla*I regions from SFFV$_P$ on an SFFV$_A$ background are being prepared (see Fig. 6). These constructs will enable us to determine the significance of the hydrophobic region of the envelope glycoprotein to the phenotype of the virus produced. In addition, site-specific mutagenesis will be carried out to actually pinpoint the exact sequence(s) that are essential for the altered Epo responsiveness induced by SFFV$_P$.

## Summary

The work reported here supports and extends previous genetic studies showing that the gene product of SFFV, gp52, is required for initiation of the erythroleukemia induced by the virus. Interestingly, there is no other known retrovirus whose structural gene product so acutely transforms hematopoietic cells in the adult mouse and in vitro. In one set of experiments, specifically set up to assess the role of the helper virus in the SFFV disease, we found that, although helper virus may facilitate leukemia induction by allowing virus spread, it is not absolutely required.

Other experiments demonstrated that, even though the LTR of other murine leukemia viruses is presumed to have a function in determining tissue specificity, the LTR of SFFV does not contribute to the erythroid nature of the SFFV disease and contains no unique sequences that are specifically required in any way. Even when it is substituted by other LTRs, including those from lymphoma-inducing retroviruses, the ensuing disease is an erythroleukemia. The difference in the role of the LTR in diseases induced by different viruses may be explained as follows. For diseases in which the LTR of the inducing virus plays a crucial function in determining tissue tropism (e.g., leukemias induced in newborns by F-MuLV and M-MuLV) the LTRs may provide efficient replication and spread of the virus in critical tissues (Chatis et al. 1983, 1984; Rosen et al. 1985). In the case of SFFV, we have shown using helper-free virus that virus spread is not necessary, and therefore that efficient virus replication, in which the LTR plays a role, is not a factor in this disease.

Finally, results were reported that deal with a more detailed look at gp52 function itself. Experiments on two substrains of SFFV have shown that it is the 3′ end of the *env* gene (p15E-related region) which determines the responsiveness of erythroid cells to erythropoietin. Because there are other intriguing aspects of this unique protein, evident from sequence analysis, such as (1) an N-terminal region encoded by endogenously acquired retroviral sequences, (2) a middle region containing fused gp70 and p15E domains that result from a large deletion in the middle of the *env* gene and (3) a truncated C-terminus (Amanuma et al. 1983; Clark and Mak 1983; Wolff et al. 1983), further studies should be directed towards increasing our understanding of the significance of these unusual properties.

## References

Amanuma H, Akiko K, Obata M, Sagata N (1983) Complete nucleotide sequence of the gene for the specific glycoprotein (gp55) of Friend spleen focus-forming virus. Proc Natl Acad Sci USA 80: 3913–3917

Axelrad AA, Steeves RA (1964) Assay for Friend leukemia virus: rapid quantitative method based on enumeration of macroscopic spleen foci in mice. Virology 24: 513–518

Bestwick RK, Boswell BA, Kabat D (1984) Molecular cloning of biologically active Rauscher spleen focus-forming virus and the sequences of its *env* gene and long terminal repeat. J Virol 51: 695–705

Chatis PA, Holland CA, Hartley JW, Rowe WP, Hopkins N (1983) Role for the 3′ end of the genome in determining disease specificity of Friend and Moloney murine leukemia. Proc Natl Acad Sci USA 80: 4408–4411

Chatis PA, Holland CA, Silver JE, Frederickson TN, Hopkins N, Hartley JW (1984) A 3′ end fragment encompassing the transcriptional enhancers of nondefective Friend virus confers erythroleukemogenicity on Moloney leukemia virus. J Virol 52: 248–254

Chung S-W, Wolff L, Ruscetti S (1987) Sequences responsible for the altered Epo responsiveness in $SFFV_P$-infected cells are localized to a 678-bp region at the 3′ end of the envelope gene. J Virol 61: 1661–1664

Clark SP, Mak TW (1982) Nucleotide sequences of the murine retrovirus Friend $SFFV_P$ long terminal repeats: identification of a structure with extensive dyad symmetry 5′ to the TATA box. Nucleic Acids Res 10: 3315–3330

Clark SP, Mak TW (1983) Complete nucleotide sequences of an infectious clone of Friend spleen focus-forming provirus: gp52 is an envelope fusion glycoprotein. Proc Natl Acad Sci USA 80: 5037–5041

Friend C (1957) Cell-free transmission in adult Swiss mice of a disease having the character of a leukemia. J Exp Med 105: 307–318

Gilboa E, Park J, Kolbe M, Hwang S, Kucherlapati R, Noonan K, Freeman H (1982) Transduction and expression of nonselectable genes using retrovirus-derived vectors. In: Gluzman Y (ed) Eukaryotic viral vectors. Cold Spring Harbor Laboratory, Cold Spring Harbor

Ishimoto A, Adachi A, Sakai K, Matsuyama M (1985) Long terminal repeat of Friend-MCF virus contains the sequence responsible for erythroid leukemia. Virology 141: 30–42

Linemeyer DL, Ruscetti SK Scolnick EM, Evans LH, Duesberg PH (1981) Biological activity of the spleen focus-forming virus is encoded by a molecularly cloned subgenomic fragment of spleen focus-forming virus DNA. Proc Natl Acad Sci USA 78: 1401–1405

Linemeyer DL, Menke JG, Ruscetti SK, Evans LH, Scolnick EM (1982) Envelope gene sequences which encode the gp52 protein of spleen focus-forming virus are required for the induction of erythroid cell proliferation. J Virol 143: 223–233

Mager DL, Mak TW, Bernstein A (1981) Quantitative colony method for tumorigenic cells transformed by two distinct strains of Friend leukemia virus. Proc Natl Acad Sci USA 781: 1703–1707

Mann R, Mulligan RC, Baltimore D (1983) Construction of a retrovirus packaging mutant and its use to produce helper-free defective retrovirus. Cell 33: 153–159

Mirand EA, Steeves RA, Ayila L, Grace JT (1968) Spleen focus formation by polycythemic strains of Friend leukemia virus. Proc Soc Exp Biol Med 127: 900–904

Rosen CA, Hazeltine WA, Lenz J, Ruprecht R, Cloyd MW (1985) Tissue selectivity of murine leukemia virus infection is determined by long terminal repeat sequences. J Virol 55: 862–866

Ruscetti S, Wolff L (1984) Spleen focus-forming virus: relationship of an altered envelope gene to the development of a rapid erythroleukemia. Curr Top Microbiol Immunol 112: 21–44

Ruscetti S, Wolff L (1985) Biological and biochemical differences between variants of spleen focus-forming virus can be localized to a region containing the 3′ end of the envelope gene. J Virol 56: 717–722

Ruscetti SK, Feild JA, Scolnick EM (1981) Polycythemia- and anaemia-inducing strains of spleen focus-forming virus differ in post-translational processing of envelope-related glycoproteins. Nature 294: 663–665

Vogt M, Haggblom C, Swift S, Haas M (1985) Envelope gene and long terminal repeat determine the different biological properties of Rauscher, Friend, and Moloney mink cell focus-inducing viruses. J Virol 55: 184–192

Weiss R, Teich N, Varmus H, Coffin J (eds) (1982) RNA tumor viruses. Cold Spring Harbor Laboratory, Cold Spring Harbor
Wendling F, Moreau-Gachelin F, Tambourin P (1981) Emergence of tumorigenic cells during the course of Friend virus leukemias. Proc Natl Acad Sci USA 78: 3614–3618
Wolff L, Ruscetti S (1985) Malignant transformation of erythroid cells in vivo by introduction of a nonreplicating retrovirus vector. Science 228: 1549–1552
Wolff L, Ruscetti S (1986) Tissue tropism of a leukemogenic murine retrovirus is determined by sequences outside of the long terminal repeats. Proc Natl Acad Sci USA 83: 3376–3380
Wolff L, Scolnick E, Ruscetti S (1983) Envelope gene of the Friend spleen focus-forming virus: deletion and insertions in 3′ gp70/p15E coding region have resulted in unique features in the primary structure of its protein product. Proc Natl Acad Sci USA 80: 4718–4722
Wolff L, Kaminchik J, Hankins WD, Ruscetti SK (1985) Sequence comparisons of the anemia- and polycythemia-inducing strains of Friend spleen focus-forming virus. J Virol 53: 570–578
Wolff L, Tambourin P, Ruscetti S (1986) Induction of the autonomous stage of transformation in erythroid cells infected with SFFV: helper virus is not required. Virology 152: 272–276

# Leukemogenesis by Transactivating Retroviruses: Bovine Leukemia Virus as a Model System*

A. Burny[1,2], E. Kettmann[1,2], M. Mammerickx[1,3], G. Marbaix[2], D. Portetelle[1,2], A. van den Broeke[2], and L. Willems[1,2]

[1] Faculty of Agronomy, 5800 Gembloux, Belgium
[2] Department of Molecular Biology, University of Brussels, 67, rue des Chevaux, Rhode-St-Genèse, Belgium
[3] National Institute for Veterinary Research, Groeselenberg 99, 1180 Brussels, Belgium

## Introduction

The most frequent malignant tumor of the bovine species is a chronic lymphatic leukemia, also called bovine malignant lymphoma or lymphosarcoma (Burny et al. 1980). Field observations and experimental transmission experiments have established the infectious character of the disease. The agent was identified as a retrovirus, exogenous to the bovine species. Until the discovery of human T lymphotropic virus (HTLV-I) and type II (HTLV-II) in 1980 and 1982 (Poiesz et al. 1980; Kalyanaraman et al. 1982), bovine leukemia virus (BLV) seemed to be a curiosity and was often referred to as a nontypical type-C virus. BLV, HTLV-I and HTLV-II have very similar genomic and structural features and obviously derive from a common ancestor (Rice et al. 1984, 1985; Sagata et al. 1984). More recently, primate viruses closely related to HTLV-I were identified (Homma et al. 1984; Miyoshi et al. 1982; Seiki et al. 1985; Yamamoto et al. 1983). The proviral DNA of these viruses contains the long terminal repeats (LTRs) and the gag (group-specific antigen), pol (polymerase and endonuclease) and env (envelope) sequences characteristic of all retroviruses. Moreover, a region called X by Seiki et al. (1983) is located between env and the 3′ LTR in HTLV-I. A similar X region is present in HTLV-II and BLV genomes. The longest open reading frame identified in X codes for a protein (tat) that transactivates the viral LTR. In the presence of tat, transcription of viral genes is greatly enhanced. It is believed that tat plays a major role in tumor induction by HTLV-I, HTLV-II or BLV. Tat-1 (tat gene of HTLV-I) codes for a 42-kd protein: tat-2 (tat gene of HTLV-II) encodes a 37-kd protein product, whilst the BLV equivalent is a 34-kd protein.

HTLV-III/LAV-1 (or HIV) (Barre-Sinoussi et al. 1983; Popovic et al. 1984), the prototype virus inducing the acquired immune deficiency syndrome (AIDS), and the more recently discovered HTLV-IV (Kanki et al. 1986) and LAV-2 (Clavel et al. 1986) are morphologically and structurally related to lentiviruses (Gonda et al. 1985; Shaw et al. 1985; Sonigo et al. 1985). Again, a simian counterpart has been

* The work performed in the authors' laboratory was assisted financially by the "Fonds Cancérologique de la Caisse Générale d'Epargne et de Retraite" and by the Ministry of Agriculture. R. Kettmann and G. Marbaix are "Maîtres de Recherche" and L. Willems is "Aspirant" of the "Fonds National de la Recherche Scientifique." A. Van den Broeke is a Fellow of the Lady Tata Memorial Trust.

discovered (Kanki et al. 1985a, b; Letvin et al. 1985), apparently more closely related to HTLV-IV/LAV-2 than to HTLV-III/LAV-1. Recent observations (S. Wain-Hobson, personal communication) indeed indicate that LAV-2 is quite remote from LAV-1. It is also significant that STLV-III does not exert pathogenic effects in its host, the African green monkey, and that HTLV-IV seems to be a weak pathogen, if at all, for man (Kanki et al. 1986).

The HTLV-I/HTLV-II/BLV and HTLV-III(LAV-1)/HTLV-IV(LAV-2)/STLV-III virus groups share the striking peculiarity to code for transactivating proteins (Derse et al. 1985; Felber et al. 1985; Goh et al. 1985; Hess et al. 1985; Lee et al. 1984; Rice et al. 1984; Rosen et al. 1985; Sagata et al. 1984, 1985; Slamon et al. 1984; Sodroski et al. 1984, 1985a, b) acting either upon transcription (HTLV-I/HTLV-II/BLV; group I viruses) or at a posttranscriptional step [HTLV-III(LAV-1)/HTLV-IV(Lav-2)/STLV-III/Visna; group II viruses]. It follows that the transactivating proteins from the group I viruses are found in the nucleus, where they interact with the viral LTR and perhaps with regulatory sequences of normal genes. The transactivating proteins from group II viruses seem to act mostly at a posttranscriptional level (Sodroski et al. 1986), perhaps regulating the mode of splicing of viral mRNAs (F. Wong-Staal, personal communication).

In this review, we shall focus on the mechanism of action of BLV as an inducer of leukemia/lymphoma in ruminants. It is widely believed that the modes of action of HTLV-I and HTLV-II are very similar and comparable to that of BLV. The AIDS-related viruses (group II above) induce increased frequencies of some neoplasms, such as Kaposi's sarcoma. Their mode of action is, however, very indirect and barely understood at the time of writing.

Infection by BLV occurs by close contact between the infected donor and the noninfected recipient. Cases of natural infection by BLV are known in domestic cattle, sheep, capybara and water buffalo (Burny et al. 1980; Marin et al. 1982). Experimentally, infection can be transmitted to goats (Oslon et al. 1981), pigs (Mammerickx et al. 1981), rabbits (C. Altaner, personal communication to D.P.), rhesus monkeys (Schödel et al. 1986) and chimpanzees (Burny et al. 1980). Body secretions and blood are the vehicles of infection, and it is overwhelmingly evident that transfer of infected cells is the major route of propagation of BLV. In numerous cases transmission is linked to husbandry and veterinary practices (stables too densely populated, repeated use of contaminated syringes and needles, etc.). In warm climates, it is possible (but has not been demonstrated experimentally) that transmission can occur through bites of hematophagous insects, especially when the donor is an animal in persistent lymphocytosis. In the latter case, successful transmission of BLV is achieved with as little as 0.1 μl blood (Mammerickx et al. 1987).

## Asymptomatic Infection

The only characteristic of asymptomatically infected animals is the presence of antibody to BLV proteins. Anti-envelope gp51 antibodies appear first, closely followed by the anti-p24 response. In general, anti-BLV antibody titer rises with time, reaching very high values (up to $10^6$) at the animal's death. BLV persists in B cells (Burny et al. 1980; Levy et al. 1987) and perhaps in other tissues. In the former targets, BLV is maintained latent due to the presence in the plasma of a repressor protein, distinct from the immunoglobulin family (Gupta and Ferrer 1982; Gupta et al. 1984). The interplay between BLV repression and BLV transactivation is an intriguing facet of BLV latency that deserves in-depth investigation. The majority of BLV-infected animals remain in good health with no significant adverse effect on production traits. In some herds, especially where animals are kept up to an advanced age, tumor cases appear. The role of the genetic make-up of the animal in onset of the tumor is presently unresolved.

## BLV Infection and Persistent Lymphocytosis

Development of persistent lymphocytosis (PL) is under the control of very poorly defined parameters, among which genetic traits are probably important. In this condition the total number of circulating lymphocytes increases spectacularly and the proportion of B cells becomes highly dominant (40%–80%) in strong contrast to the low B:T ratio (1:4) prevailing in the normal individual (Burny et al. 1980; Fossum et al. in preparation). Furthermore, the number of blastoid B lymphocytes is increased up to 30-fold in the blood from BLV-infected animals (Fossum et al. in preparation). In general, PL is rather stable and may last for years. It does not seem, however, to be a necessary step toward development of the tumor. The tumor stage can break out without any obvious change in the number of circulating lymphocytes. Among PL lymphocytes, about 30% carry the BLV provirus; the remaining 70% are reactive cells (the majority of the B type) with no trace of BLV information. As mentioned above, repression of BLV expression depends upon the presence of a nonimmunoglobulin protein present in the blood plasma. Short-term culture of BLV provirus-carrier lymphocytes in the absence of the repressor or transfer of such lymphocytes to a naive recipient results in the lifting of repression and induction of virus expression. Allogenic lymphocytes will produce virus in the contaminated host until they will be rejected by the immune system.

## BLV Infection and the Tumor Phase

A fraction only (0.1%–10%) of infected recipients will develop the tumor phase during the normal lifespan of a domestic animal. The major properties of tumor cells are as follows:

1. All tumors contain at least a fragment of the provirus. This suggests that some viral functions are strictly required for initiation of the neoplastic process.
2. Tumors are made of an homogeneous population of cells harboring up to four copies of the provirus. Tumors are monoclonal with respect to the integration site of the provirus (Kettmann et al. 1980). Integration sites, however, are not conserved from one animal to the other.

3. Whether unique or multiple, proviral copies are complete or deleted. All deleted copies examined have shown preservation of the tat region, stressing again its probable essential role in the tumoral process.
4. BLV information is totally or almost completely silent, even in tumor cells grown in vitro (Kettmann et al. 1985). The six cell lines that we have developed so far have, however, not been cloned. Therefore we cannot decide whether the slight BLV expression is due to some contaminating nontransformed cells or to leakage of the leukemic block in some transformed cells leading to cell differentiation and virus expression.
5. The apparent lack of expression of viral functions suggests that virus expression is necessary to initiate the neoplastic process. Once a stage of no return has been reached, the virus can be switched off.
6. As the virus replicates in B cells, it is customary to consider that tumor cells are of the B lineage. None of our six cell lines, however, expresses immunoglobulins, either in the culture medium or at the cell surface or in the cytoplasm. Two bovine tumor cell lines show very high expression at the cell surface of class II molecules of the major histocompatibility complex. Obviously, availability and use of very specific reagents is mandatory in order to unravel the identity of the tumor cell.

The mode of cell transformation by BLV, HTLV-I and HTLV-II remains conjectural. Beyond any doubt, the viral tat gene plays a key role within a given cellular background. The virus-cell interplay leads the cell to a given stage where it can stay for ever being blocked in its differentiation pathway. It is certain that the virus is necessary but by no means sufficient. Given sets of circumstances rarely encountered, or rare secondary events, must contribute to push the cell across barriers beyond which an irreversible step toward full malignancy has been taken. We know nothing about the nature and number of circumstances and events involved in leading a cell to transformation.

Considering that the virus is mandatory in the initial steps of the process and apparently dispensable later on, we hypothesize that the transient expression of viral functions can lead to permanent expression of critical cellular genes. We foresee two possible explanations:

1. Expression of the viral tat protein induces with very low frequency (development of the tumor being a very rare event) expression of a cell protein that positively modulates its own production and represses production of the *tat* protein. In the absence of tat, all viral functions are suppressed. The cell protein stimulates its own production in an autocatalytic way. Similar situations (positive regulatory loops) have been demonstrated in the bacteriophage $\lambda$ system, for example.
2. Expression of the viral tat protein induces chromosomal abnormalities that definitely stabilize the transformed stage. Examples of chromosomal rearrangements in neoplasia are numerous, such as the translocation of the Philadelphia chromosome in CML (Heisterkamp et al. 1983), the various translocations affecting the myc oncogene in Burkitt lymphoma (Taub et al. 1982) and the chromosome breaks bcl1 and bcl2 in B cell lymphomas (Tsujimoto et al. 1984a, b). Taking as an example a situation recently described in yeast (Meeks-Wagner and Hartwell 1986), we could imagine that tat expression leads with low frequency to an imbal-

ance of histone class proteins which in turn induces chromosomal abnormalities. Evidently, many biochemical pathways could be susceptible to deregulation by the tat protein of HTLV-I/HTLV-II/BLV. The initial event of transformation by these viruses has been uncovered and the road to better understanding of cell transformation is wide open.

## Summary

Bovine leukemia virus (BLV) is the etiological agent of a chronic lymphatic leukemia in cows, sheep and goats. Human T lymphotropic virus type I (HTLV-I) induces a T cell leukemia, and its type II counterpart has been found in dermatopathic lymphadenopathy, hairy cell leukemia and prolymphocytic leukemia cases. Moreover, HTLV-I and II are apparently involved in neurological diseases of the degenerative type.

BLV, HTLV-I and HTLV-II share many genomic and structural characteristics. Their most striking feature is the existence of an X region located between the *env* gene and the long terminal repeat sequence. The X region encompasses several genes, among which *tat* (for transactivator of transcription) is now identified. The *tat* product probably plays a role in the induction process of the tumor phase. BLV system is an adequate model for studies of cell-virus interplay leading to transformation.

## References

Barre-Sinoussi F, Chermann JC, Rey F, Nugeyre MT, Chamaret S, Gruest J, Dauguet C, Axler-Blin C, Vezinet-Brun F, Rouzioux C, Rozenbaum W, Montagnier L (1983) Isolation of a T lymphotropic retrovirus from a patient at risk for acquired immune deficiency syndrome (AIDS). Science 220: 868–870

Burny A, Bruck C, Chantrenne H, Cleuter Y, Dekegel D, Ghysdael J, Kettmann R, Leclercq M, Leunen J, Mammerickx M, Portetelle D (1980) Bovine leukemia virus: molecular biology and epidemiology. In: Klein G (ed) Viral oncology. Raven, New York, pp 231–289

Clavel F, Guetard D, Brun-Vezinet F, Chamaret S, Rey MA, Santos-Ferreira MO, Laurent AG, Dauguet C, Katlama C, Rouzioux C, Klatzmann D, Champalimaud JL, Montagnier L (1986) Isolation of a new human retrovirus from West African patients with AIDS. Science 233: 343–346

Derse D, Caradonna SJ, Casey JW (1985) Bovine leukemia virus long terminal repeat. A cell type-specific promoter. Science 227: 317–320

Felber B, Paskalis H, Kleinman-Ening C, Wong-Staal F, Pavlakis GN (1985) The px protein of HTLV-I is a transcriptional activator of its long terminal repeats. Science 229: 675–677

Goh WC, Sodroski J, Rosen C, Essex M, Haseltine WA (1985) Subcellular localization of the product of the long open reading frame of human T-cell leukemia virus type 1. Science 227: 1227–1228

Gonda M, Wong-Staal F, Gallo RC, Clements JE, Narayan O, Gilden RV (1985) Sequence homology and morphology similarity of HTLV-III and visna virus, a pathogenic lentivirus. Science 227: 173–175

Gupta P, Ferrer JF (1982) Expression of bovine leukemia virus genome is blocked by a non-immunoglobulin protein in plasma from infected cattle. Science 215: 405–407

Gupta P, Kashmiri SVS, Ferrer JF (1984) Transcriptional control of the bovine leukemia virus ge-

nome: role and characterization of a non-immunoglobulin plasma protein from bovine leukemia virus infected cattle. J Virol 50: 267–270
Heisterkamp N, Stephenson JR, Groffen J, Hansen PF, de Klein A, Bartram CR, Grosveld G (1983) Localization of the c-abl oncogene adjacent to a translocation break point in chronic myeloïd leukaemia. Nature 306: 239–242
Hess JL, Clements JE, Narayan O (1985) Cis- and trans-acting transcriptional regulation of visna virus. Science 229: 482–485
Homma T, Kanki PJ, King NM, Hunt RD Jr, O'Connell MJ, Letvin NL, Daniel MD, Desrosiers RC, Yang CS, Essex M (1984) Lymphoma in macaques: association with virus of human T lymphotropic family. Science 225: 716–718
Kalyanaraman VS, Sarngadharan MG, Robert-Guroff M, Miyoshi I, Blayney D, Golde D, Gallo RC (1982) A new subtype of human T-cell leukemia virus (HTLV-II) associated with a T-cell variant of hairy cell leukemia. Science 218: 571–573
Kanki PJ, Kurth R, Becker W, Dreesman G, McLane MF, Essex M (1985a) Antibodies to simian T-lymphotropic retroviruses type III in African green monkeys and recognition of STLV-II viral proteins of AIDS and related sera. Lancet I: 1330–1332
Kanki PJ, McLane MF, King NW, Letvin NL, Hunt RD, Seghal P, Daniel MD, Desrosiers RC, Essex M (1985b) Serologic identification and characterization of a macaque T-lymphotropic retrovirus closely related to HTLV-III. Science 228: 1199–1201
Kanki PJ, Barin F, M'Boup S, Allan JS, Romet-Lemonne JL, Marlink R, McLane MF, Lee TH, Arbeille B, Denis F, Essex M (1986) New human T-lymphotropic retroviruses related to simian T-lymphotropic virus type III ($STLV-III_{AGM}$). Science 232: 238–240
Kettmann R, Cleuter Y, Mammerickx M, Meunier-Rotival M, Bernardi G, Burny A, Chantrenne H (1980) Genomic integration of bovine leukemia provirus: comparison of persistent lymphocytosis with lymph node tumor form of enzootic bovine leukosis. Proc Natl Acad Sci USA 77: 2577–2581
Kettmann R, Cleuter Y, Grégoire D, Burny A (1985) Role of the 3′ long open reading frame region of bovine leukemia virus in the maintenance of cell transformation. J Virol 54: 899–901
Lee TH, Coligan JE, Sodroski JG, Haseltine WA, Salahuddin JZ, Wong-Staal F, Gallo RC, Essex M (1984) Antigens encoded by the 3′-terminal region of human T-cell leukemia virus: evidence for a functional gene. Science 226: 57–61
Letvin NL, Daniel MD, Sehgal PK, Desrosiers RC, Hunt RD, Waldron LM, McKey JJ, Schmidt DK, Chalifoux LV, King NW (1985) Introduction of AIDS-like disease in macaque monkeys with T-cell tropic retrovirus STLV-III. Science 230: 71–73
Levy D, Kettmann R, Marchand P, Djilali S, Parodi AL (1987) Selective tropism of bovine leukemia virus (BLV) for surface immunoglobulin-bearing ovine B lymphocytes. Leukemia 1: 463–465
Mammerickx M, Portetelle D, Burny A (1981) Experimental cross-transmission of bovine leukemia virus (BLV) between several animal species. Zentralbl Veterinarmed [B] 28: 69–81
Mammerickx M, Portetelle D, De Clercq C, Burny A (1987) Experimental transmission of bovine leukemia virus to cattle, sheep and goats: infectious doses of blood and incubation period of the infection. Leuk Res 11: 353–358
Marin C, de Lopez N, de Alvarez L, Castanos H, Espana W, Leon A, Bello A (1982) Humoral spontaneous response to bovine leukemia virus infection in zebra, sheep, buffalo, and capybara. Curr Top Vet Med Animal Sci 15: 310–320
Meeks-Wagner D, Hartwell LH (1986) Normal stoichiometry of histone dimer sets is necessary for high fidelity of mitotic chromosome transmission. Cell 44: 43–52
Miyoshi I, Yoshimoto S, Fujishita M, Taguchi H, Kubonishi I, Niiya K, Minezawa M (1982) Natural adult T-cell leukaemia virus infection in Japanese monkeys. Lancet II: 658
Olson C, Kettmann R, Burny A, Kaja R (1981) Goat lymphosarcoma from bovine leukemia virus. JNCI 67: 671–675
Poiesz BJ, Ruscetti FW, Gazdar AF, Bunn PA, Minna JD, Gallo RC (1980) Isolation of type C retrovirus particles from cultured and fresh lymphocytes of a patient with cutaneous T-cell lymphoma.
Popovic M, Sarngadharan MG, Read E, Gallo RC (1984) Detection, isolation and continuous production of cytopathic retroviruses (HTLV-III) from patients with AIDS and pre-AIDS. Science 224: 497–500
Rice NR, Stephens RM, Couez D, Deschamps J, Kettmann R, Burny A, Gilden RV (1984) The nucleotide sequence of the env gene and post-env region of bovine leukemia virus. Virology 138: 82–93

Rice NR, Stephens RM, Burny A, Gilden RV (1985) The gag and pol genes of bovine leukemia virus: nucleotide sequence and analysis. Virology 142: 357–377

Rosen CA, Sodroski JG, Kettmann R, Burny A, Haseltine WA (1985) Trans activation of the bovine leukemia virus long terminal repeat in BLV-infected cell. Science 227: 320–322

Sagata N, Yasunaga T, Ohishi K, Tsuzuku-Kawamura J, Onuma M, Ikawa Y (1984) Comparison of the entire genomes of bovine leukemia virus and human T-cell leukemia virus and characterization of their unidentified open reading frames. EMBO J 3: 3231–3237

Sagata N, Tsuzuku-Kawamura J, Nagaioshi-Aida M, Shimizu F, Imagawa KI, Ikawa Y (1985) Identification and some biochemical properties of the major X-BL gene product of bovine leukemia virus. Proc Natl Acad Sci USA 82: 7879–7882

Schödel F, Hahn B, Hübner R, Hochstein-Mintzel V (1986) Transmission of bovine leukemia virus (BLV) to immunocompromised monkeys: evidence for persistent infection. Microbiologica 9: 163–172

Seiki M, Hattori S, Hirayama Y, Yoshida M (1983) Human adult T-cell leukemia virus: complete nucleotide sequence of the provirus genome integrated in leukemia cells DNA. Proc Natl Acad Sci USA 80: 3618–3622

Seiki M, Watanabe T, Komuro A, Miyoshi I, Hayami M, Yoshida M (1985) Characterization of simian retrovirus genome related to human T-cell leukemia virus type 1. In: Retroviruses in human lymphoma/leukemia. Miwa M, Sugano H, Sugimura T, Weiss RA (eds) Japan Scientific Societies Press, Tokyo and VNU Science Press, Utrecht, pp 241–249

Shaw GM, Harper ME, Hahn BH, Epstein LG, Gajduzek DC, Price RN, Navia BA, Petito CK, O'Hara CJ, Groopman JE, Cho ES, Oleske JM, Wong-Staal F, Gallo RC (1985) HTLV-III infection in brains of children and adults with AIDS encephalopathy. Science 227: 177–182

Slamon DJ, Shimotohno K, Cline MJ, Golde DW, Chen ISY (1984) Identification of the putative transforming protein of the human T-cell leukemia viruses HTLV-I and HTLV-II. Science 226: 61–65

Sodroski JG, Rosen CA, Haseltine WA (1984) Trans-acting transcriptional activation of the long terminal repeat of human T lymphotropic viruses in infected cells. Science 225: 381–385

Sodroski J, Rosen C, Goh WC, Haseltine WA (1985a) A transcriptional activator protein encoded by the x-lor region of the human T-cell leukemia virus. Science 228: 1430–1432

Sodroski J, Rosen C, Wong-Staal F, Salahuddin SZ, Popovic M, Arya S, Gallo RC, Haseltine WA (1985b) Trans-acting transcriptional regulation of human T-cell leukemia virus type III long terminal repeat. Science 227: 171–173

Sodroski J, Goh WC, Rosen C, Dayton A, Terwilliger E, Haseltine WA (1986) A second post-transcriptional trans-activator gene required for HTLV-III replication. Nature 321: 412–417

Sonigo D, Alizon M, Staskus K, Klatzmann D, Cole S, Danos O, Retzel E, Tiollais P, Haase A, Wain-Hobson S (1985) Nucleotide sequence of the visna lentivirus: relationship to the AIDS virus. Cell 42: 369–382

Taub R, Kirsch I, Morton C, Lenoir G, Swan D, Tronick S, Aaronson S, Leder P (1982) Translocation of the c-myc gene into the immunoglobulin heavy chain locus in human Burkitt lymphoma and murine plasmacytoma cells. Proc Natl Acad Sci USA 79: 7837–7841

Tsujimoto Y, Yunis J, Onorato-Showe L, Erikson J, Nowell PC, Croce C (1984a) Molecular cloning of the chromosomal break point of B-cell lymphomas and leukemias with the t (11; 14) chromosome translocation. Science 224: 1403–1406

Tsujimoto Y, Finger LR, Yunis J, Nowell PC, Croce C (1984b) Cloning of the chromosome break point of neoplastic B-cells with the t (14; 18) chromosome translocation. Science 226: 1097–1099

Yamamoto N, Hinuma Y, zur Hausen H, Schneider J, Hunsmann G (1983) African green monkeys are infected with adult T-cell leukemia virus or a closely related agent. Lancet I: 240–241

# The Structure and Function of Reovirus Genes*

Wolfgang K. Joklik

Department of Microbiology and Immunology, Duke University Medical Center, Durham, NC 27710, USA

## Introduction

Reoviruses possess two capsid shells, each composed of morphological subunits arranged according to icosohedral symmetry principles. The genome of reoviruses consists of 10 segments of double-stranded (ds) RNA. Upon infection these molecules are transcribed into plus-stranded single-stranded (ss) RNA by an RNA polymerase, the transcriptase, that is present within reovirus cores. These transcripts fulfill two functions: they can be translated into the reovirus proteins, and they can be transcribed into the minus strands with which they remain associated, thereby forming progeny dsRNA molecules. Thus they act as the reovirus messenger RNAs, and they also represent the physical as well as the genetic link between parental and progeny reovirus genomes.

The purpose of this paper is to examine current knowledge concerning the structure of the reovirus genome segments and the nature and function of the proteins that they encode. The following considerations are pertinent in this regard. First, it is clear that each of the 10 reovirus messenger RNA (mRNA) molecules is translated into a major protein species the size of which approximates the size of the protein expected to be encoded by each mRNA molecule. The total number of reovirus-encoded proteins is greater, however, for two reasons.

1. Several of the reovirus proteins are cleaved to provide cleavage products which in all probability also possess a function. For example, the primary translation product of the m2 species of reovirus mRNA is a protein, μ1, which is a minor core component; but more than 90% of protein μ1 is cleaved to protein μ1C, the major component of the reovirus outer capsid shell (Smith et al. 1969). In fact, protein μ1C is the major protein component of reovirus particles. The fact that protein μ1 is cleaved to protein μ1C was first shown by pulse chase experiments which clearly demonstrated the time-dependent conversion (Zweerink and Joklik 1970); by the fact that monoclonal antibodies against protein μ1 also precipitate μ1C (Lee et al. 1981a); by V8 peptide map analysis (McCrae and Joklik 1978);

* This work was supported by Research Grant 1R01 AI08909, Program Project Grant 2P01 CA30246 and Research Training Grants 5T32 AI07148 and 2T32 CA09111 from the National Institutes of Health.

and very recently the gene that encodes protein μ1 has been sequenced and the location of the cleavage site identified by determination of the N-terminal amino acid sequence of protein μ1C, which indicates that the cleavage of μ1 to μ1C occurs within the N-terminal portion of the μ1 molecule (Fields, personal communication).

Another example is provided by the nonstructural protein μNS, which exists in cells in roughly equal amounts in uncleaved form and as protein μNSC, a cleavage product about 5 kd smaller (Lee et al. 1981 a). Finally, reovirus particles always contain small amounts of a protein that is precipitated by monoclonal antibodies against protein λ2 and that is therefore a cleavage product of this major core component (Lee et al. 1981 a).

2. The second reason why there are more reovirus proteins than genome segments is the fact that some of the reovirus mRNAs appear to be polycistronic. In particular, mRNA species s1 encodes not only the primary protein σ1, translation of which is initiated at the first AUG, but also the much smaller protein σ1S, which is translated in a different reading frame (Nagata et al. 1984; Ernst and Shatkin 1985; Cashdollar et al. 1985; Jacobs and Samuel 1985; Sarkar et al. 1985). Protein σ1S is made in quite large amounts and is a small basic protein about 120 amino acids long. Other reovirus genome segments have also been sequenced and some of them also contain short open reading frames that may also give rise to small proteins.

The second consideration arises from an examination of the nature of the functions that are expected to be performed by reovirus-encoded proteins. About half of them appear to be structural components of reovirus particles. However, other functions are known to be performed by reovirus proteins and can presently be grouped into two classes. The first is made up of enzymes. Reovirus cores transcribe the 10 genome segments into mRNA molecules that are capped, but not polyadenylated. Available evidence indicates that the capping enzymes are also components of reovirus cores. These capping enzymes comprise a nucleotide phosphohydrolase that hydrolyzes 5′-terminal ppp groups to 5′-pp groups; a guanylylating enzyme; and two methylases, for methylating the cap G and the ribose residue that was the 5′-terminal ribose of uncapped transcripts respectively. The nature of these enzymes is not known because reovirus cores are very stable structures; all attempts to disrupt cores in order to examine their components for enzymic activity results in total loss of activity.

Second, it is known that there are two primary reovirus proteins that are not structural proteins; these are proteins μNS and σNS. These two nonstructural proteins are made in large amounts in infected cells, and available evidence indicates that they perform some essential function during the first stages of reovirus morphogenesis, when sets of the 10 individual plus strands are assembled into complexes within which they are transcribed into minus strands with which they remain associated, thereby providing the progeny genomes. These immature reovirus particles then mature, via a series of protein exchange reactions, first into core-like particles and finally into mature reovirus particles.

The function of viral proteins can be examined by a variety of techniques, among which are the following. First, the proteins can be isolated and their properties examined. As will be discussed below, we are in the process of placing cloned reovirus genome segments into expression vectors in order to provide sufficient amounts of the native primary reovirus proteins for studies of their properties and functions. Second, one can compare the reovirus multiplication cycles in cells infected at nonpermissive temperatures with wild-type virus and with temperature-sensitive mutants. Temperature-sensitive *(ts)* mutants in all 10 genome segments are available and such studies are indeed underway. Third, one can compare multiplication cycles in cells infected with variants of reovirus that differ in the functional properties of one particular genome segment or of the protein that it encodes. This approach is also feasible for reovirus, since mammalian reoviruses belong to one of three serotypes. As I will show below, the serotype-specific protein is protein $\sigma 1$, encoded by genome segment S1, and each of the three allelic forms of genome segment S1 is associated in nature with its own distinct set of nine genome segments. Therefore there exist in nature three forms of each of the 10 genome segments. Examination of multiplication cycles in cells infected with variants that differ in one genome segment, the phenotype of which can be experimentally distinguished, therefore provides another way in which the nature of viral protein function can be investigated.

My purpose here is to illustrate these three approaches in attempting to identify the nature of the functions exercised by the various reovirus-encoded proteins.

## Genetic Relatedness of the Three Reovirus Serotypes

The extent of relatedness among the three reovirus serotypes can be measured in a variety of ways. The simplest is to measure the extent to which the genomes of each of the three serotypes can hybridize to those of the other two serotypes under standardized conditions (Gaillard and Joklik 1982). Table 1 shows the results. Serotypes 1 and 3 are related to the extent of about 70% and serotype 2 is related to serotypes 1 and 3 to the extent of about 10%. It is possible to estimate the extent of relatedness among the cognate genome segments by constructing hybrid genome segments in which one strand is that of one serotype and the other is that of another. Then one can again measure the proportion of unpaired regions by RNase susceptibility, using standard conditions (that is, when homologous hybrids yield no RNase-digestible material and when 100% of ssRNA is digested). When this analysis is carried out for all genome segments, the results shown in Table 2 are obtained (Gaillard and Joklik 1982). The most dissimilar segments are the S1 segments, none of which are more than 10% related to each other. This is not surprising, because, as pointed out above, the protein encoded by the S1 genome segment, protein $\sigma 1$, is the reovirus type-specific antigen. Another noteworthy feature of this analysis is that all genome segments of serotype 2 are far less related to those of serotypes 1 and 3 than those of 1 and 3 are to each other. This supports the statement made above, namely that the S1 genome segment of each serotype is associated in nature with its own set of the other nine genome segments; in other words, the three reovirus serotypes evolved as homologous gene sets.

**Table 1.** Measurement of total genome homology among reovirus serotypes 1, 2, and 3

| | $^3$H-Labeled plus-stranded RNA | | |
|---|---|---|---|
| Genome dsRNA | Serotype 1 | Serotype 2 | Serotype 3 |
| Serotype 1 | 100 | 10 | 73 |
| Serotype 2 | 10 | 100 | 11 |
| Serotype 3 | 72 | 8 | 100 |

From Gaillard and Joklik (1982)
Data expressed as adjusted and corrected percentages of ribonuclease-resistant material

**Table 2.** Percentage homology among the genome segments of reovirus serotypes 1, 2, and 3

| Genome segment | +(1)[a] / −(2) | +(1) / −(3) | +(3) / −(1) | +(3) / −(2) | +(2) / −(1) | +(2) / −(3) |
|---|---|---|---|---|---|---|
| L1 | 15 | 86 | 84 | 18 | 21 | 11 |
| L2 | 23 | 85 | 94 | 13 | 12 | 7 |
| L3 | 12 | 88 | 79 | 13 | 15 | 14 |
| M1 | 9 | 66 | 69 | 5 | 8 | 3 |
| M2 | 5 | 31 | 42 | 9 | 11 | 8 |
| M3 | 10 | 63 | 67 | 5 | 6 | 8 |
| S1 | 4 | 12 | 6 | 1 | 1 | 4 |
| S2 | 14 | 56 | 44 | 9 | 6 | 13 |
| S3 | 12 | 69 | 81 | 14 | 8 | 12 |
| S4 | 11 | 57 | 51 | 11 | 11 | 15 |

From Gaillard and Joklik (1982)
[a] Percentage ribonuclease-resistant material in the hybrid genome segment consisting of a serotype 1 plus strand and a serotype 2 minus strand

## The Sequences at the Termini of Reovirus Genome Segments

The fact that the genome segments of serotype 2 are so distantly related to those of serotypes 1 and 3 was surprising in view of the fact that RNA sequence analysis of the two terminal regions of several genome segments of all three serotypes had indicated that such terminal regions are very closely related (Gaillard et al. 1982). Figure 1 shows the 5′-terminal regions of the genome segments L3, M3, S1, and S2 of all three serotypes. Clearly the S1 genome segments are very dissimilar, but the 5′-terminal regions of the other three genome segments are related more closely than predicted by the hybridization analysis illustrated in Table 2.

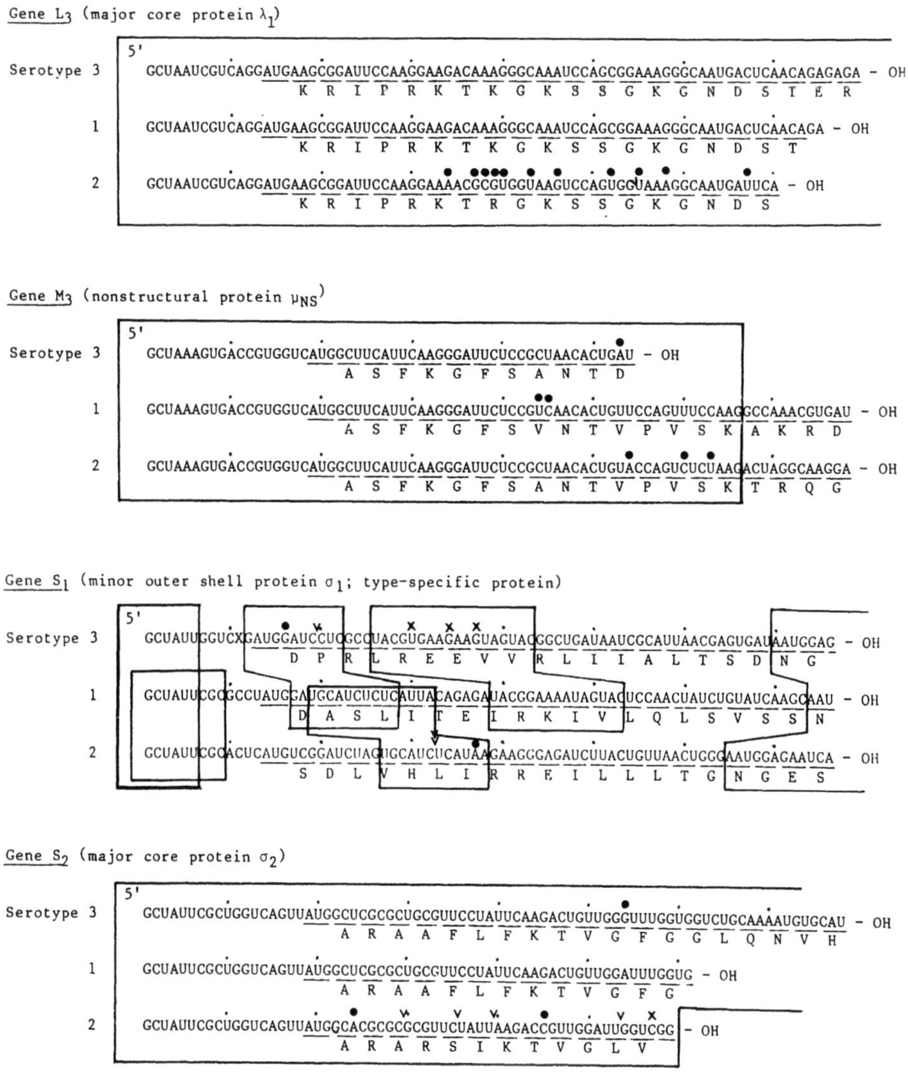


**Fig. 1.** The sequences of the 5′-terminal regions of the plus strands of genome segments L3, M3, S1, and S2 of reovirus serotypes 1, 2, and 3. The sequences shown here are the complements of the actually sequenced 3′-terminal regions of the corresponding minus strands. Base changes are indicated by ● above the line (the odd base out of three being indicated); deletions (bases missing) are indicated by v; and insertions (bases inserted into the sequence of one genome segment, but not into that of the other two) are indicated by ×. (From Gaillard et al. 1982)

## Serologic Relatedness of the Proteins Encoded by the Three Reovirus Serotypes

Another indication along these lines came from the results of an immunologic analysis of the relatedness of the major epitopes of reovirus proteins. This analysis was carried out as follows (Gaillard and Joklik 1980). Cells were infected with each of the three reovirus serotypes and the proteins that were made from 10.5 to 12 h after infection were labeled with [$^{35}$S]-methionine. The cell lysates were then reacted with various dilutions of antisera against each of the three reovirus serotypes, and the antigen-antibody complexes were collected on Staph A immunoadsorbent. Following polyacrylamide gel electrophoresis and autoradiography, for each protein the antiserum concentration was determined that was capable of precipitating 50% of the maximum amount of protein that could be precipitated. The reciprocals of these concentrations are recorded in Table 3. It is seen that the antigenic determinants on most cognate proteins were very similar, that is, they were able to react equally well with all three antisera. There were only a few proteins with significantly different epitopes (differing by more than ± twofold). The only protein with very significantly different epitopes was protein σ1. In fact, serotype 3 σ1 protein exhibited absolute

**Table 3.** Quantitation of the serologic relatedness of reovirus-coded proteins

| Protein | Coded by serotype | Antiserum against | | |
|---|---|---|---|---|
| | | Serotype 1 | Serotype 2 | Serotype 3 |
| λ1 | 1 | 250 | 180 | 160 |
| | 2 | 190 | 200 | 240 |
| | 3 | 110 | 140 | 120 |
| λ2 | 1 | 230 | 60 | 120 |
| | 2 | 260 | 160 | 200 |
| | 3 | 80 | 240 | 1000 |
| μ1 | 1 | 660 | 140 | 520 |
| | 2 | 400 | 200 | 660 |
| | 3 | 200 | 410 | 700 |
| μ1C | 1 | 1420 | 80 | 700 |
| | 2 | 1000 | 1100 | 1220 |
| | 3 | 700 | 1780 | 1100 |
| μNS | 1 | 40 | 30 | 30 |
| | 2 | 40 | 40 | 30 |
| | 3 | 60 | 30 | 40 |
| σ1 | 1 | 1240 | 60 | <10 |
| | 2 | 100 | 200 | <10 |
| | 3 | <10 | <10 | 2000 |
| σ2 | 1 | 280 | 100 | 1100 |
| | 2 | 280 | 1250 | 2300 |
| | 3 | 320 | 400 | 820 |
| σ3 | 1 | 1000 | 260 | 1060 |
| | 2 | 1060 | 2400 | 1700 |
| | 3 | 860 | 1660 | 680 |
| σNS | 1 | 40 | 40 | 45 |
| | 2 | 30 | 45 | 30 |
| | 3 | 50 | 40 | 50 |

From Gaillard and Joklik (1980)
Data expressed as reciprocals of the antiserum dilutions at which 50% of the maximum amount of protein is precipitated. *Boxes* indicate significant type specificity

type specificity; that is, it was precipitated 100 times less efficiently by antisera against serotypes 1 and 2 than by antiserum against serotype 3. Type specificity was less absolute for the serotype 1 and 3 σ1 proteins, but it was nevertheless highly significant.

The most remarkable feature of this analysis was the fact that in so many cases serotype 2 proteins were precipitated as efficiently by antisera against serotypes 1 and 3 as by homologous antiserum, considering the results shown in Table 1, which indicate that the genetic relatedness of the corresponding genome segments is no more than 10%–20%. Thus, the sequences that encode the epitopes of most reovirus genome segments tend to be highly conserved.

## The Sequences of Reovirus Genome Segments

All 10 genome segments of reovirus serotype 3 have been cloned by a process that involved polyadenylating both plus and minus strands of genome segments, transcribing the separated segments with reverse transcriptase, digesting RNA strands with alkali and annealing the transcribed cDNA strands. The dsDNA molecules were then C-tailed and cloned into the *Pst*I site of pBR322 (Cashdollar et al. 1982). Figure 2 demonstrates that all ten genome segments were cloned. Both ends of all cloned genome segments were then sequenced and the sequences compared with the actual sequences of the terminal regions as determined by rapid RNA sequencing. It was shown in this manner that all cloned genome segments were intact copies of the naturally occurring genome segments.

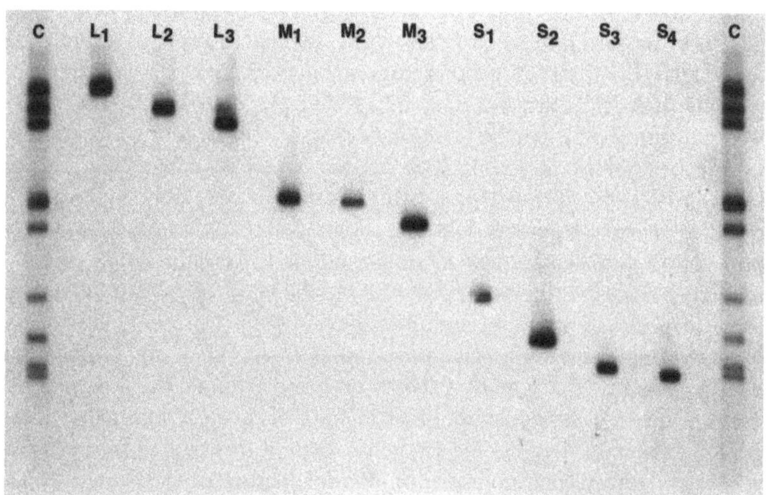


**Fig. 2.** Identification of cloned reovirus genes. Total reovirus dsRNA was electrophoresed in a 7.5% polyacrylamide gel. The RNA was transferred to diazotized aminophenylthioether paper and hybridized to total $^{125}$I-labeled reovirus dsRNA to localize the genes (lanes C) and to various cloned DNAs labeled with $^{32}$P by nick translation (lanes L1–S4). (From Cashdollar et al. 1984)

Several of the cloned genome segments have by now been sequenced; they are genome segments S1 (Nagata et al. 1984; Cashdollar et al. 1985; Bassel-Duby et al. 1985), S2 (Cashdollar et al. 1982), S3 (Richardson and Furuichi 1983), and S4 (Giantini et al. 1984). Currently we are in the process of sequencing segments L2 and L3, as well as M1 and M3, and other laboratories are sequencing genome segment M2 (Fields, personal communication). The sequences of genome segments S2, S3, and S4 exhibit no unusual features; their sequences have been compared with all sequences in the Protein Sequence Database of the Protein Identification Resource of the National Biomedical Research Foundation, and no significant similarities have been noted.

## The Sequence of the Three S1 Genome Segments

An analysis of particular interest was to sequence all three S1 genome segments. These are the genome segments that have diverged most extensively during evolution, most probably because of the physical location of protein $\sigma1$. Protein $\sigma1$ is present in reovirus particles to the extent of 48 molecules, 4 at each of the 12 icosahedral vertices (Lee et al. 1981b). It is an outer protein coat component, but it does not appear to be part of a capsomer. It may therefore be under fewer structural constraints than other structural proteins that are capsomer components; and this may be the reason why it has evolved much more rapidly than other outer capsid proteins in response to immunologic pressures. However, in spite of this marked divergence, the three $\sigma1$ proteins are still found at the same locations in reovirus particles, they are still the type-specific antigens, and they still serve as the reovirus cell attachment proteins. The reason for sequencing the three S1 genome segments was to identify the characteristics of the evolutionary divergence process. Figure 3 shows the sequences of the three S1 genome segments (Cashdollar et al. 1985; Nagata et al. 1984; Bassel-Duby et al. 1985). The sequences present several interesting features. First, the three $\sigma1$ proteins are encoded by the open reading frames that start with the first AUG at residue 13 or 14. The three $\sigma1$ proteins are 399–455 codons long; an interesting feature is the fact that the serotype 3 $\sigma1$ protein is the largest, yet electrophoretic migration data suggest that it is smaller than the $\sigma1$ proteins of serotypes 1 and 2. The reason for this anomalous behavior is not known. Second, all three S1 genome segments possess a short open reading frame that starts at an initiation codon at about residue 70 and is about 120 codons long. These second open reading frames overlap those that encode the $\sigma1$ proteins and all are terminated by three in-phase termination codons. These short $\sigma1$ proteins, the $\sigma1S$ proteins, are basic proteins, and their function is not known. They are synthesized in infected cells in amounts that greatly exceed, on a molar basis, the amounts of $\sigma1$ proteins that are made. It is remarkable that in spite of the very marked evolutionary divergence of the three S1 genome segments, these three reading frames have been highly conserved with respect to initiation codons, nature of the proteins transcribed, and the three in-phase termination codons.

In distinction to proteins encoded by genome segments S2, S3, and S4, which show no similarity to any of the proteins presently in the Protein Sequence Database, the 3 $\sigma1$ proteins show strong homology with eukaryotic myosin and tropo-

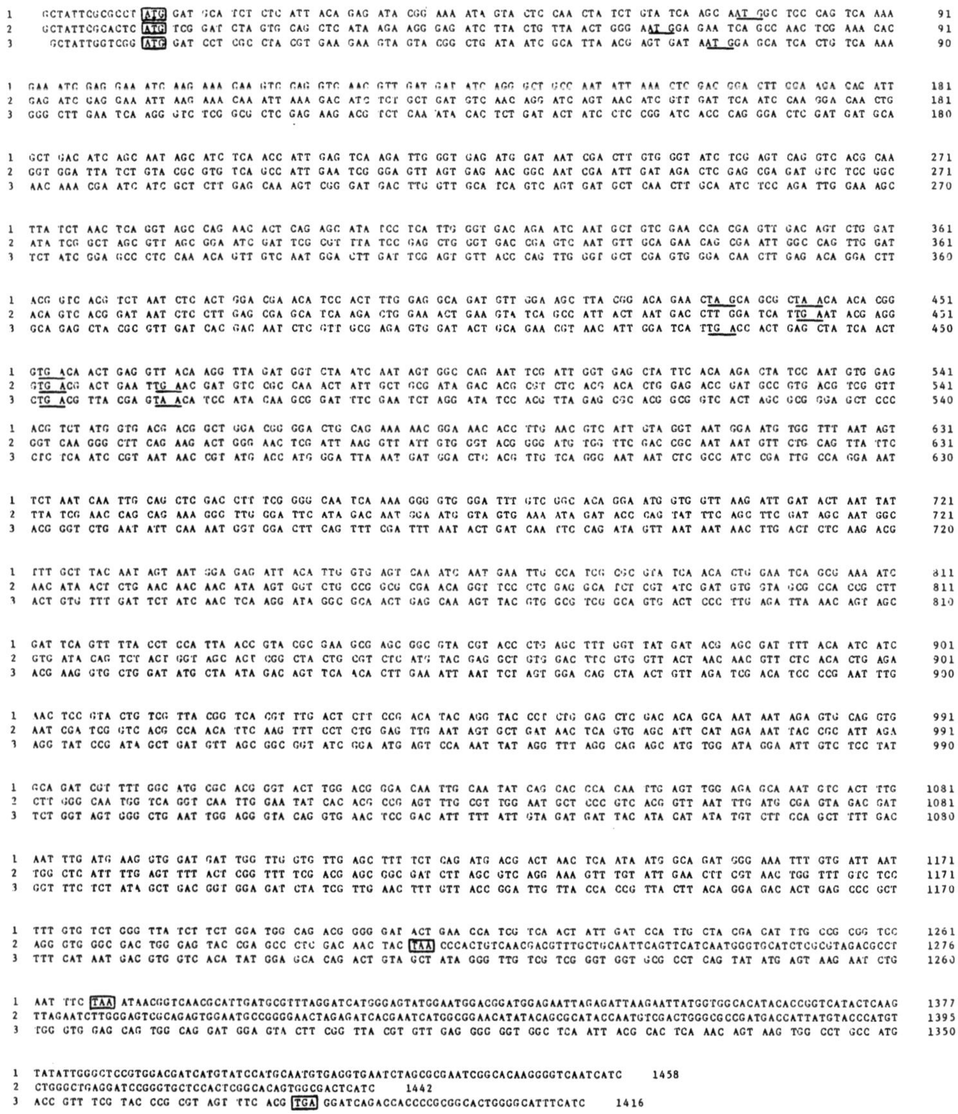

**Fig. 3.** Sequences of the serotype 1–3 S1 genome segments of reovirus. The sequences are those of the plus (mRNA) strands. (From Cashdollar et al. 1985)

myosin (Bassel-Duby et al. 1985). It is conceivable, therefore, that, like these proteins, the tetramers of $\sigma 1$ molecules in reovirus particles exist in the form of $\alpha$-helices wound around each other in a coiled-coil structure that is inserted into the reovirus projections or spikes. The C-terminal domain of protein $\sigma 1$ would then be postulated to project out from the virus particles in a globular structure that contains both the cell-attachment site and the epitope (Bassel-Duby et al. 1985).

The genetic relatedness at the RNA level of the coding regions of the three S1 genome segments is 28%, 5%, and 9% respectively for serotypes 1:2, 1:3, and 2:3; the corresponding relatedness values of the three σ1 proteins are 36%, 16%, and 14%. The sequences of the three σ1 proteins are shown in Figure 4 (Cashdollar et al. 1985).

A fascinating feature of this analysis is that the S1 genome segments and σ1 proteins of serotypes 1 and 2 are related much more closely to each other than to those of serotype 3. In other words, whereas for nine genome segments there is a serotype 1:3 relatedness pattern, for the S1 genome segments and σ1 proteins it is a serotype 1:2 relatedness pattern. An indication that this might be so was provided by the earlier finding that the serotype 3 σ1 protein exhibited absolute species specificity, whereas the σ1 proteins of serotype 1 and 2 showed merely high, but not absolute, species specificity.

The most likely explanation of this crossover relatedness pattern is that at some relatively recent stage during evolution, exchange of serotype 2 and 3 S1 genome segments (via genome segment reassortment) occurred, favored perhaps by increased heterologous capsid stability resulting from acquisition by protein σ1 of a new structural feature, or the characteristics of the then prevalent immunologic pressures, perhaps modified by acquisition of a new host species.

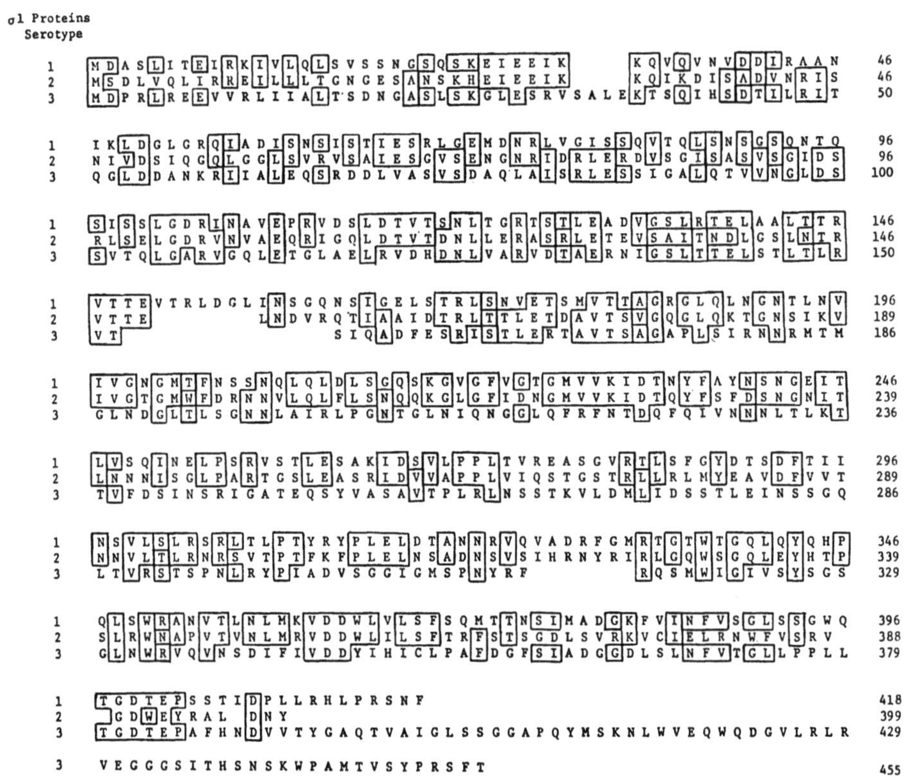


**Fig. 4.** Sequences of the σ1 proteins of reovirus serotypes 1–3. (From Cashdollar et al. 1985)

## The Translation Efficiencies of the Various Reovirus Messenger RNAs

The reovirus system provides an excellent model system for studying the features that control efficiency of mRNA translation; for in reovirus-infected cells 10 species of mRNA that are synthesized in roughly equimolar amounts are translated with vastly different efficiencies, and all reovirus mRNAs and proteins can be quantitated with considerable accuracy. As a result, it is possible to measure the efficiencies with which the various species of reovirus mRNA are translated (Gaillard and Joklik 1985). Figure 5 shows the results normalized with respect to the mRNA species that is translated most efficiently at any particular time, and the actual translation efficiencies are presented in Table 4. It is clear that there is an enormous range of translation efficiencies, a range that exceeds 100-fold. The most efficiently translated mRNA species is usually species s4, followed by species m2, which, on average, is translated about two-thirds as efficiently. Species s2 and s3 are translated slightly less than one-half as efficiently as species s4; species m3, l2 and l3 one-quarter to one-third as efficiently; and species s1 one-tenth as efficiently. Finally, species l1 and m1 are translated very inefficiently under all conditions; their translation efficiencies are usually no more than 1% of that of species s4. We are now modifying the sequences upstream of and surrounding the initiation codons of several of the mRNA species by gene splicing techniques in order to examine in depth the nature and characteristics of the sequences that control translation efficiency.

## The Functions of the Nonstructural Proteins μNS and σNS

We have been carrying out an in vivo analysis of the morphogenesis of reovirus. Our approach has been to synchronize morphogenesis either by infecting cells with a high multiplicity of wild-type virus or with a *ts* mutant at the nonpermissive temperature followed by shiftdown, and then to react cell lysates with monoclonal antibodies against various reovirus proteins. The aim of these experiments is to find monoclonal antibodies that will precipitate all 10 species of ssRNA; that is, antibodies that are directed against proteins with which all 10 species of ssRNA are associated. Presumably these proteins would be the components of immature reovirus particles during the initial stages of reovirus morphogenesis. This approach works very well, and using it we have found that the two proteins that are the first to combine with all 10 species of ssRNA are the two nonstructural proteins σNS and μNS. Evidently these two nonstructural proteins function during the early stages of reovirus mor-

**Table 4.** Relative translation efficiencies of reovirus mRNAs

| Species | l1 | l2 | l3 | m1 | m2 | m3 | s1 | s2 | s3 | s4 |
|---|---|---|---|---|---|---|---|---|---|---|
| | <2 | 27 | 23 | <3 | 64 | 31 | 8 | 45 | 39 | 96 |

Data expressed as the averages for all three reovirus serotypes, throughout multiplication cycles at both 38 °C and 33 °C. (Gaillard and Joklik 1985)

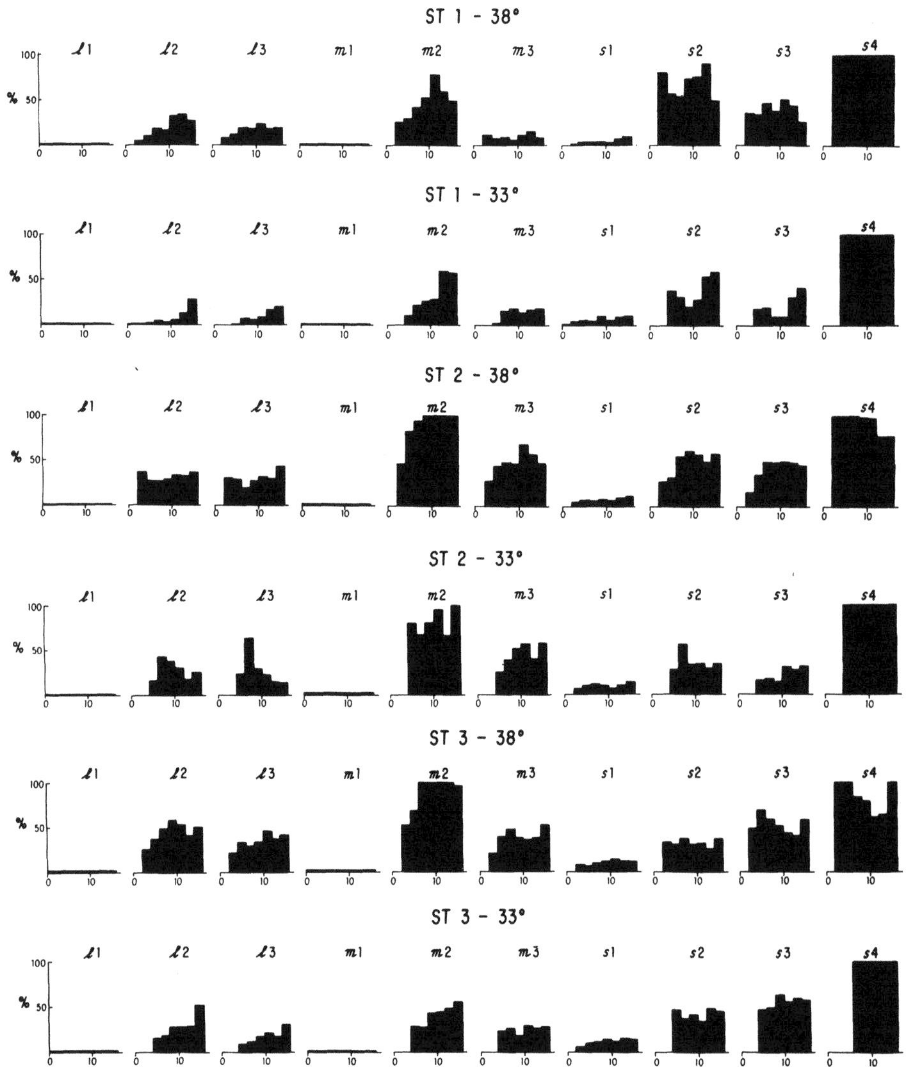


**Fig. 5.** Histograms depicting the translation efficiencies of the individual species of reovirus serotype 1, 2, and 3 mRNAs at 38 °C and 33 °C at various stages of the multiplication cycle. Autoradiograms were prepared of polyacrylamide gels and the amounts of radioactive label present in bands corresponding to each species of reovirus ssRNA and protein were determined by laser densitometry. The ratios of the amount of radioactive label incorporated into each protein species during each 2-h period divided by the amount of radioactive label present in each cognate mRNA species at that time were then calculated and expressed as percentages of the value for the mRNA species for which this ratio was highest (i.e., that which was translated most efficiently). (From Gaillard and Joklik 1985)

phogenesis. The sequence of protein σNS is already available (Richardson and Furuichi 1983); it shows no significant similarity to any other protein in the Protein Sequence Database. We have almost finished sequencing genome segment M3, which encodes protein μNS.

## Placing Reovirus Genes into Expression Vectors

Two reovirus genome segments have so far been cloned into expression vectors in which the proteins that they encode are expressed in amounts sufficient for functional studies to be carried out. First, genome segment S3 has been expressed in pRC23 under the control of the $\lambda P_L$ promoter to yield protein σNS that has been purified and found to possess affinity for ssRNA, like authentic protein σNS (Richardson and Furuichi 1985). Interestingly, this binding was inhibited by GTP. Second, protein σ1 has been expressed under the control of the lac promoter by pUC13 as a fusion protein 14 amino acids longer than authentic σ1 (Masri et al. 1986). Both the host cell attachment domain and the hemagglutinin domain, as well as the antigenic sites, of this fusion protein were indistinguishable from those of authentic protein σ1. We are engaged in a major program placing all cloned reovirus genome segments into expression vector systems, with the purpose of isolating sufficient amounts of the native proteins to isolate them and determine whether they possess enzymic activity, RNA or protein binding ability, or some other demonstrable function. Further, the availability of adequate amounts of native reovirus proteins should permit in vitro experiments on reovirus morphogenesis, experiments that should complement the studies on in vivo morphogenesis described above.

We are employing both prokaryotic and eukaryotic expression systems for this purpose. One of the prokaryotic expression vectors that we are using is pDR540, the tac promoter of which causes the S2 genome segment to be expressed in quite large amounts. We are also using the baculovirus expression system and are expressing protein σ1S in it in large amounts. Finally, we are exploring several new mammalian expression vector systems, in particular one based on a strong cowpox virus promoter.

## Enzymic Activities of Reovirus Proteins

It should be pointed out that two reovirus proteins have already been implicated as possessing enzymic activity on the basis of indirect but convincing evidence. These are protein λ2, which may be the guanylyl transferase (Shatkin et al. 1983), and protein λ3, which may be a transcriptase component because it controls the pH optimum of the transcriptase (Drayna and Fields 1982).

In summary, within a short period of time all reovirus genome segments will be sequenced and computer analyses may be able to suggest the functions of at least some of the reovirus proteins. Reovirus genome segments are also being placed into expression vectors and will soon be available in amounts adequate for direct functional studies. Finally, reovirus *ts* mutants and reassortants are available for analysis of the reovirus multiplication cycle under conditions when a single reovirus protein is either incapable of functioning or functions in a detectably different manner. These analyses will also point the way to the actual functions of the various reovirus proteins.

## References

Bassel-Duby R, Jayasuriya A, Chatterjee D, Sonenberg N, Maizel JV Jr et al (1985) Sequence of reovirus hemagglutinin predicts an alpha helical region and shows homology to myosin. Nature 315: 421–423

Cashdollar LW, Esparza J, Hudson GR, Chmelo RA, Lee PWK et al (1982) Cloning the double-stranded RNA genes of reovirus: sequence of the cloned S2 gene. Proc Natl Acad Sci USA 79: 7644–7648

Cashdollar LW, Chmelo RA, Esparza J, Hudson GR, Joklik WK (1984) Molecular cloning of the complete genome of reovirus serotype 3. Virology 133: 191–196

Cashdollar LW, Chmelo RA, Wiener JR, Joklik WK (1985) Sequences of the S1 genes of the three serotypes of reovirus. Proc Natl Acad Sci USA 82: 24–28

Drayna D, Fields BN (1982) Activation and characterization of the reovirus transcriptase: genetic analysis. J Virol 41: 110–118

Ernst H, Shatkin AJ (1985) Reovirus hemagglutinin mRNA codes for two polypeptides in overlapping reading frames. Proc Natl Acad Sci USA 82: 48–52

Gaillard RK, Joklik WK (1980) The antigenic determinants of most of the proteins coded by the three serotypes of reovirus are highly conserved during evolution. Virology 107: 533–536

Gaillard RK, Joklik WK (1982) Quantitation of the relatedness of reovirus serotypes 1, 2, and 3 at the gene level. Virology 123: 152–164

Gaillard RK, Joklik WK (1985) The relative translation efficiencies of reovirus messenger RNAs. Virology 147: 336–348

Gaillard RK, Li JK-K, Keene JD, Joklik WK (1982) The sequences at the termini of four genes of the three reovirus serotypes. Virology 121: 320–326

Giantini M, Seliger LS, Furuichi Y, Shatkin AJ (1984) Reovirus type 3 genome segment S4: nucleotide sequence of the gene encoding a major virion surface protein. J Virol 52: 984–987

Jacobs BL, Samuel CE (1985) Biosynthesis of reovirus-specified polypeptides: the reovirus s1 mRNA encodes two primary translation products. Virology 143: 63–74

Lee PWK, Hayes EC, Joklik WK (1981a) Characterization of anti-reovirus immunoglobulins secreted by cloned hybridoma cell lines. Virology 108: 134–146

Lee PWK, Hayes EC, Joklik WK (1981b) Protein σ-1 is the reovirus cell attachment protein. Virology 108: 156–163

Masri SA, Nagata L, Mah DCW, Lee PWK (1986) Functional expression in *Escherichia coli* of cloned reovirus S1 gene encoding the viral cell attachment protein σ1. Virology 149: 83–90

McCrae MA, Joklik WK (1978) The nature of the polypeptide encoded by each of the ten double-stranded RNA segments of reovirus type 3. Virology 89: 578–593

Nagata L, Masri SA, Mah DCW, Lee PWK (1984) Molecular cloning and sequencing of the reovirus (serotype 3) S1 gene which encodes the viral cell attachment protein σ1. Nucleic Acids Res 12: 8699–8710

Richardson MA, Furuichi Y (1983) Nucleotide sequence of reovirus genome segment S3, encoding non-structural protein σNS. Nucleic Acids Res 11: 6399–6408

Richardson MA, Furuichi Y (1985) Synthesis in *Escherichia coli* of the reovirus nonstructural protein σNS. J Virol 56: 527–533

Sarkar G, Pelletier J, Bassel-Duby R, Jayasuriya A, Fields BN, Sonenberg N (1985) Identification of a new polypeptide coded by reovirus gene S1. J Virol 54: 720–725

Shatkin AJ, Furuichi Y, LaFiandra AJ, Yamakawa M (1983) Initiation of mRNA synthesis and 5′-terminal modification of reovirus transcripts. In: Compans RW, Bishop DHL (eds) Proc Symp Double-Stranded RNA Viruses. Elsevier Biomedical, New York, pp 43–54

Smith RE, Zweerink HJ, Joklik WK (1969) Polypeptide components of virions, top component, and cores of reovirus type 3. Virology 39: 791–810
Zweerink HJ, Joklik WK (1970) Studies on the intracellular synthesis of reovirus-specified proteins. Virology 41: 501–522

# Selective Inhibition of Virus Replication

HANS J. EGGERS

Institut für Virologie der Universität zu Köln, Fürst-Pückler-Strasse 56, 5000 Köln 41, Federal Republic of Germany

Despite great strides in animal virology since 1949 it was maintained for a long period of time that selective inhibition of virus multiplication, and consequently antiviral chemotherapy, would be impossible. One main argument was based on the obligate intracellular parasitism of viruses, i. e., the intimate dependence of virus replication on the metabolism of the host cell. It was further argued that even if suitable antiviral agents were available, their medical application would come too late, since with appearance of the first symptoms of disease virus multiplication had already reached its peak. Thus, even in the face of the discovery of the interferon system, those interested in antiviral chemotherapy remained a small group of outsiders.

Our working hypothesis, though still vague in the beginning, was as follows: Virus replication is a highly regulated series of processes involving virus-specific enzymes and effector molecules, and not a straightforward, simple replication of RNA or DNA and protein molecules, ultimately embracing each other to form new viral particles. There exist, in the virus-infected cell, virus-specific molecules and structures which are not present in the uninfected cell. Selective compounds may recognize and combine with these virus-specific structures, rendering them unable to function (Tamm and Eggers 1963; Eggers and Tamm 1966). Viral targets might be virus-induced nonstructural molecules, structural parts of the virion, or the virus particle itself.

In the past some advances along these lines have been made, and, most importantly, well-founded, promising results in the treatment of virus diseases in man can now be recorded.

Basic immunology tells us that a virus particle is recognized by the organism as a foreign structure, leading to the formation of virus-specific antibodies and often ensuing in neutralization of viral infectivity. Perhaps an analogous mechanism operates with low-molecular weight substances. Some time ago, we described the antiviral activity of rhodanine (2-thio-4-oxothiazolidine) on the replication of echovirus 12 (Eggers 1970; Eggers et al. 1970). Other enteroviruses or members of other groups of viruses appeared unaffected. Also, virus-inhibitory concentrations of rhodanine had no obvious effects on uninfected cells. Studies on the mechanism of antiviral activity of rhodanine revealed that the compound affected exclusively an early step of virus replication, viz. uncoating of echovirus 12 (Eggers 1970, 1977). Uncoating was measured as acquisition of light resistance of neutral red-sensitized virus in infected cells. It was further shown that rhodanine, at virus-inhibitory concentrations, protects specifically the echo 12 virion from heat or alkaline degradation. Poliovirus, for instance, is not stabilized by the compound. These experimental data are compatible with the hypothesis that rhodanine interacts with echovirus 12-specific surface structures, stabilizing the particle so that the viral genome cannot be released for successful initiation of virus replication. It should be noted that in the presence of rhodanine other properties of the virus particle remain unaffected. For example, echovirus 12 hemagglutinates in the presence of rhodanine to high titer, and the virus particle adsorbs to and enters the host cell (Eggers 1977). Electron micrographs and biologic experiments reveal that the virion containing all structural polypeptides, resides in the cytoplasm as an inert particle (Eggers et al. 1979; Rosenwirth and Eggers 1979).

The same principle, viz. inhibition of virus multiplication via inhibition of uncoating, appears to operate in the case of arildone (4-[6-(2-chloro-4-methoxyphenoxy)hexyl]-3,5-heptanedione) and all three serological types of poliovirus (Mc Sharry et al. 1979). As far as picornaviruses are concerned, the antiviral activity of arildone seems to be restricted to polioviruses, and poliovirus particles are stabilized by the compound against inactivation by heat or alkaline pH (McSharry et al. 1979).

In contrast to rhodanine-echovirus 12, the arildone-poliovirus system can be tested in an experimental animal, since poliovirus type 2 infects mice and leads, as in primates, to neurogenic paralysis (Eggers et al. 1983; Roskopf 1985). In first experiments, the toxicity of arildone was evaluated in newborn and 3-week-old mice. Administration of 1250 mg arildone/kg body weight subcutaneously twice a day for 10 days was well tolerated; in particular, weight gain was unaffected. Higher doses could not be tested due to insolubility of the compound.

In the actual experiment, newborn mice were infected intracerebrally with about 800 infective doses per mouse, and treatment was begun at the same time by inoculation of 62.5 mg arildone/kg body weight subcutaneously twice a day for 20 days. Untreated control animals developed paresis about 2–3 days post virus inoculation and all had died by 1 week p.i. from poliomyelitis. Arildone-treated mice, on the other hand, remained well up to day 14 p.i., but all had died by 3 weeks after virus infection. This outcome could not be improved substantially by increasing the dose of arildone or lowering the virus dose inoculated.

This experiment allows discussion of an important problem in antiviral chemotherapy. Why did the animals ultimately die from poliomyelitis though treatment had been ongoing? One explanation could be the emergence of drug-resistant or drug-dependent virus mutants, a problem first recognized with picornavirus-selective inhibitors such as benzimidazole derivatives or guanidine in the early 1960s (Eggers and Tamm 1961, 1962b, 1963b). It could be shown, however, that in this case the cause of the final therapeutic failure was not emergence of arildone-resistant virus mutants, but rather maturation of metabolic activities, which by the 12$^{th}$ day of life had reached an extent such that effective levels of arildone could no longer be maintained. The virus now multiplied uninhibited, and after 3–4 days, around day 15 or 16 p.i., poliomyelitis became manifest (Roskopf 1985).

A single injection of poliovirus antiserum on day 10 or day 12 p.i. after initial treatment with arildone, however, prevented poliomyelitis permanently (Roskopf 1985).

The foregoing experiments illustrate that a substance might be effective and of high selectivity in cell culture, but a failure when administered in the organism. This possibility had led us to perform the tests in newborn animals.

Another more general consideration concerns the start of arildone injections in relation to the time of virus infection in our experiment. Obviously, we exercised hemoprophylaxis rather than antiviral chemotherapy. Consequently, setting aside the problem of metabolic inactivation, could a substance like arildone ever be used therapeutically in humans infected with poliovirus, or would treatment come too late, as initially maintained by some investigators? It is obvious that the experimental conditions in our mouse experiment do not in any way reflect natural infection and pathogenesis of poliomyelitis in man: poliovirus is inoculated directly into the central nervous system of the mouse, and we have shown that within 12–24 h p.i. the virus reaches titers of $10^8$ plaque-forming units per gram brain or spinal cord. It is established, however, that in man poliovirus multiplies first in non-nervous tissue, mostly producing symptoms, before possibly ultimately invading the central nervous system.

In passing, it may be mentioned that poliovirus also replicates in the heart of experimental animals, causing a diffuse myocarditis with muscle cell degeneration and adjacent invasion of inflammatory cells (Eggers et al. 1983). The pathogenesis of myocarditis in patients with poliomyelitis has remained unknown, since multiplication of virus in the heart has never been proven unequivocally.

Besides rhodanine and arildone, two other antiviral agents apparently act on the early stages of virus replication, viz. amantadine (1-aminoadamantane) and rimantadine ($\alpha$-methyl-1-adamantanemethylamine) (Davies et al. 1964; Kato and Eggers 1969). Both of these compounds selectively inhibit the replication of influenza A viruses, and are effective in the chemoprophylaxis and chemotherapy of influenza A in man. The molecular basis of the virus-inhibitory activity of amantadine appears complex. It has been known for a long time that inhibition of influenza A virus replication extends over a wide range of concentrations of amantadine. At concentrations of $>0.1$ m*M* the compound may inhibit activation of the membrane fusion activity of the virus hemagglutinin, resulting from an elevation in the pH of endosomes (Hay et al. 1985). At lower concentrations the primary target of antiviral action seems to be the $M_2$ membrane protein. The $M_2$ protein is thought to play a role in initiation of virus infection, but also in virus assembly (Hay et al. 1985). Hay et al. (1985) consider amantadine "an example of a drug targeted specifically against a viral membrane protein," a statement which underscores the general concept of antiviral chemotherapy proposed many years ago.

Several controlled studies on the efficacy of amantadine and rimantadine in the prophylaxis and early treatment of influenza A have been carried out. As an example a double-blind, randomized trial of the prophylactic effects of rimantadine and amantadine during an outbreak of influenza A (mostly H1N1 subtype) in Burlington, Vermont, may be cited (Dolin et al. 1982). The volunteers received drugs or placebo orally at a dose of 100 mg twice a day for 6 weeks. It was found (Table 1) that rimantadine provided an efficacy rate [(rate of illness in placebo group minus rate in rimantadine (or amantadine) group/rate in placebo group) × 100] of 85% in persons with laboratory-documented influenza. Amantadine yielded an efficacy rate of 91%. The efficacy rates in cases with influenza-like illness (i.e., without laboratory confirmation of influenza virus infection) were somewhat lower, as expected.

**Table 1.** Influenza-like illness and laboratory-documented influenza among volunteers receiving placebo, rimantadine, or amantadine

| Treatment group | Influenza-like illness (%) | Laboratory-documented influenza (%) |
|---|---|---|
| Placebo ($n = 132$) | 41 | 21 |
| Rimantadine ($n = 133$) | 14 | 3 |
| Efficacy rate | 65 | 85 |
| Amantadine ($n = 113$) | 9 | 2 |
| Efficacy rate | 78 | 91 |

From Dolin et al. 1982

An important issue ever since the drugs have been administered to patients or healthy persons has been their side effects. The side effects in the present study concerned essentially the central nervous system, with symptoms like insomnia, "jitteriness," and difficulty in concentrating. More recipients of amantadine (13%) than of rimantadine (6%) or placebo (4%) withdrew from the study because of central nervous system side effects. It is for this reason that rimantadine appears to be the drug of choice for the prophylaxis of influenza.

Though for high-risk persons primarily vaccination against influenza is recommended, it is obvious that many fail to be vaccinated in time. Besides, the efficacy of vaccination is only about 70%, not counting the possiblity of a new influenza A virus variant.

In the case of rhodanine and arildone, it seems reasonable to assume that these selective inhibitors interact with structures of the virus particle (Eggers and Rosenwirth 1988). However, in the virus-infected cell, nonstructural proteins may also provide targets for virus-selective inhibitors. Picornaviruses, for instance, require for multiplication an enzyme, RNA-dependent RNA polymerase, capable of elongating nascent RNA chains from an RNA template, an activity not present in the uninfected cell (Baltimore et al. 1963). It has long been known that certain benzimidazole derivatives inhibit specifically the synthesis of enterovirus RNA (Eggers and Tamm 1962a, 1963a) and thereby virus replication. The primary target of the antiviral action is not known: the RNA-dependent RNA polymerase, for example, as assumed originally (Baltimore et al. 1963), or a translation product of the P2 region. It is clear, however, that with addition of the compounds viral RNA synthesis (Eggers 1982), and consequently production of infectious virus (Halperen et al. 1964), is stopped immediately. Only formation of empty capsids is ongoing, for about 2 h (Halperen et al. 1964).

Several of the principal studies on antiviral chemotherapy have been carried out with benzimidazole derivatives, for instance those investigating the problem of drug-resistance and drug-dependence (Eggers and Tamm 1961, 1962b, 1963b; Eggers et al. 1963). It was shown early in this system that mutation frequencies in the order of $10^{-4}$ mutations per replication are not uncommon (Eggers and Tamm 1965). Nevertheless, in an enterovirus mouse model, the emergence of drug-resistant mutants played no significant role, since the drug-resistant virus mutants investigated had lost virulence to an appreciable extent (Eggers 1976). It remains to be seen whether this observation is of more general significance.

Unfortunately, benzimidazoles, like arildone, are readily metabolized to inactive compounds, thus preventing their use in primates (Eggers 1982; Federmann 1980; Hecker 1981). All attempts to circumvent oxidation or glucuronidation of the compounds by designing alternative structures have been unsuccessful so far. It could be shown in the mouse system, however, that animals already paralyzed by echovirus 9 infection could still be treated successfully (Eggers 1982).

It needs hardly be stated that certain virus species carry in the virus particle an enzyme required for replication. Of obvious immediate interest is the reverse transcriptase of LAV/HTLV III, the etiologic agent of the acquired immune deficiency syndrome. Despite worldwide efforts to find a selective inhibitor of this enzyme (or other virus-specific functions), a breakthrough is not in sight. However, the administration of azidothymidine (AZT) to patients with AIDS has been shown to provide significant beneficial effects.

Significant advances have been made in recent years in the treatment of herpesvirus infections in man, notably herpes simplex and varicella-zoster virus infections. In particular, immunocompromised patients (leukemia, neoplasias, antitumor treatment, transplantations) are highly vulnerable to severe and life-threatening infections with these viruses.

With 9-(2-hydroxyethoxymethyl)guanine(acycloguanosine, acyclovir) an effective and apparently well-tolerated compound is at hand (Elion et al. 1977; Schaeffer et al. 1978). In the herpesvirus-infected cell this acyclic analog of guanosine is readily converted to the 5′-monophosphate by the virus-induced deoxypyrimidine kinase complex, whereas phosphorylation is restricted in uninfected cells. After formation of acyclo-GMP, it is finally converted by cellular kinases to the 5′-triphosphate, acyclo-GTP, the active metabolite. Acyclo-GTP selectively inhibits the herpesvirus-induced DNA polymerase. In consequence, herpesvirus DNA synthesis is specifically inhibited. In addition, acyclo-GTP is incorporated into viral DNA, which obviously leads to chain termination, since the molecule lacks the 3′ hydroxyl group. As a result, with acyclovir a number of significant differences between herpesvirus-infected and uninfected cells are exploited, which may readily explain the high antiviral potency and selectivity of the compound.

Besides for immunocompromised patients, acyclovir has been used successfully for the treatment of herpes keratitis and of primary herpes genitalis. In some patients with frequently recurring and severe herpes genitalis, long-term prophylactic treatment with acyclovir has been tried with considerable success. In this situation the possibility to administer the compound orally is of great advantage. Unfortunately, however, recurrences appear immediately when the prophylactic treatment is discontinued. This outcome was to be expected, since there is no basis for an effect of acyclovir on herpesvirus in its latent state.

Finally, the significant therapeutic effects of acyclovir on herpesvirus encephalitis should be mentioned. Herpesvirus encephalitis is a severe disease with a lethality of about 70%. The majority of survivors suffer from neuropsychiatric sequelae. In pioneering studies, Whitley et al. (1977 and 1981) have shown that administration of 9-$\beta$-D-arabinofuranosyladenine (ara-A, vidarabine) will significantly improve the outcome of this severe infection (Table 2). This has been confirmed in subsequent investigations. A further improvement has been achieved with acyclovir, evaluated versus vidarabine in biopsy-proven or serologically proven cases of herpes encepha-

**Table 2.** Treatment of herpesencephalitis: vidarabine vs acyclovir

| | Vidarabine[a] (%) | | Acyclovir[b] (%) | | Authors |
|---|---|---|---|---|---|
| Survival | 60.2 | ($n=93$) | – | | Whitley et al. (1977, 1981) |
| Restitution | 32.3 | | – | | |
| Survival | 50.0 | ($n=24$) | 81.5 | ($n=27$) | Sköldenberg et al. (1984) |
| Restitution | 12.5 | | 55.5 | | |
| Survival | 48.6 | ($n=37$) | 81.2 | ($n=32$) | Whitley et al. (1986) |
| Restitution | 13.5 | | 37.5 | | |

[a] 15 mg/kg/day for 10 days [b] 10 mg/kg 3 times a day for 10 days

litis (Table 2). An overall survival rate of about 80% and overall restitution of about 45% represent a major achievement in the face of a disease with such a severe prognosis. The outcome, as one would expect, is positively influenced by early treatment. Another significant factor is the age of the patients (Sköldenberg et al. 1984; Whitley et al. 1986).

In this chapter, only some aspects of antiviral chemotherapy could be discussed. It is obvious that many problems remain to be solved. A more efficient screening system for promising selective inhibitors is needed, not to mention guidelines for a rational design of such compounds. The advances in elucidation of the three-dimensional structure of viral macromolecules may have a substantial impact on antiviral chemotherapy (Smith et al. 1986). Nevertheless, significant advances have been made in this field, and chemotherapy of viral diseases in man is now feasible.

## References


Baltimore D, Eggers HJ, Franklin RM, Tamm I (1963) Poliovirus-induced RNA polymerase and the effects of virus-specific inhibitors on its production. Proc Natl Acad Sci USA 49: 843–849

Davies WL, Grunert RR, Haff RF, McGahen JW, Neumayer EM, Paulshock M, Watts JC, Wood TR, Hermann EC, Hoffmann CE (1964) Antiviral activity of 1-adamantanamine (amantadine). Science 144: 862–863

Dolin R, Reichman RC, Madore HP, Maynard R, Linton PN, Weber-Jones J (1982) A controlled trial of amantadine and rimantadine in the prophylaxis of influenza A infection. N Engl J Med 307: 580–584

Eggers HJ (1970) Inhibition of early stages of virus-cell interactions. Ann NY Acad Sci 173: 417–419

Eggers HJ (1976) Successful treatment of enterovirus-infected mice by 2-($\alpha$-hydroxybenzyl)-benzimidazole and guanidine. J Exp Med 143: 1367–1381

Eggers HJ (1977) Selective inhibition of uncoating of echovirus 12 by rhodanine. Virology 78: 241–252

Eggers HJ (1982) Benzimidazoles: selective inhibitors of picornavirus replication in cell cultures and in the organism. In: Came PE, Caliguiri LA (eds) Chemotherapy of viral infections. Springer, Berlin Heidelberg New York, pp 377–417 (Handbook of experimental pharmacology, vol 61)

Eggers HJ, Tamm I (1961) Spectrum and characteristics of the virus inhibitory action of 2-($\alpha$-hydroxybenzyl)-benzimidazole. J Exp Med 113: 657–682

Eggers HJ, Tamm I (1962 a) On the mechanism of selective inhibition of enterovirus multiplication by 2-($\alpha$-hydroxybenzyl)-benzimidazole. Virology 18: 426–438

Eggers HJ, Tamm I (1962b) A variant of coxsackie A9 virus which requires 2-(α-hydroxybenzyl)-benzimidazole (HBB) for optimal growth (Abstr). 8th Int Congr Microbiol, p 85
Eggers HJ, Tamm I (1963a) Inhibition of enterovirus ribonucleic acid synthesis by 2-(α-hydroxybenzyl)-benzimidazole. Nature 197: 1327–1328
Eggers HJ, Tamm I (1963b) Drug dependence of enteroviruses: variants of coxsackie A9 and ECHO 13 viruses that require 2-(α-hydroxybenzyl)-benzimidazole for growth. Virology 20: 62–74
Eggers HJ, Tamm I (1965) Coxsackie A9 virus: mutation from drug dependence to drug independence. Science 148: 97–98
Eggers HJ, Tamm I (1966) Antiviral chemotherapy. Annu Rev Pharmacol 6: 231–250
Eggers HJ, Reich E, Tamm I (1963) The drug-requiring phase in the growth of drug-dependent enteroviruses. Proc Natl Acad Sci USA 50: 183–190
Eggers HJ, Koch MA, Furst A, Daves GD Jr, Wilczynski JJ, Folkers K (1970) Rhodanine: a selective inhibitor of the multiplication of echovirus 12. Science 167: 294–297
Eggers HJ, Bode B, Brown DT (1979) Cytoplasmic localization of the uncoating of picornaviruses. Virology 92: 211–218
Eggers HJ, Roskopf U, Arnold G (1983) Myocarditis in experimental poliomyelitis. A contribution to the pathogenesis of the disease. Zentralbl Bakteriol Hyg 1 Abt Orig A 255: 164
Eggers HJ, Rosenwirth B (1988) Isolation and characterization of an arildone-resistant poliovirus 2 mutant with an altered capsid protein VP1. Antiviral Res 9: 23–26
Elion GB, Furman PA, Fyfe JA, DeMiranda P, Beauchamp L, Schaeffer HJ (1977) Selectivity of action of an antiherpetic agent, 9-(2-hydroxyethoxymethyl) guanine. Proc Natl Acad Sci USA 74: 5716–5720
Federmann G (1980) Behandlung von Enterovirus-Infektionen der saugenden Maus mit D-HBB · HCl und chemischen Derivaten. Dissertation, University of Cologne, Cologne
Halperen S, Eggers HJ, Tamm I (1964) Evidence for uncoupled synthesis of viral RNA and viral capsids. Virology 24: 36–46
Hay AJ, Wolstenholme AJ, Skehel JJ, Smith MH (1985) The molecular basis of the specific anti-influenza action of amantadine. EMBO J 4: 3021–3024
Hecker C (1981) Therapieversuche bei Enterovirusinfektionen in neugeborenen Mäusen. Dissertation, University of Cologne, Cologne
Kato N, Eggers HJ (1969) Inhibition of uncoating of fowl plague virus by 1-adamantanamine hydrochloride. Virology 37: 632–641
McSharry JJ, Caliguiri LA, Eggers HJ (1979) Inhibition of uncoating of poliovirus by arildone, a new antiviral drug. Virology 97: 307–315
Rosenwirth B, Eggers HJ (1979) Early processes of echovirus 12-infection: elution, penetration, and uncoating under the influence of rhodanine. Virology 97: 241–255
Roskopf U (1985) Studien zur Pathologie und Chemotherapie der Poliomyelitis in der Maus. Dissertation, University of Bonn, Bonn
Schaeffer HJ, Beauchamp L, DeMiranda P., Elion GB, Bauer DJ, Collins P (1978) 9-(2-hydroxyethoxymethyl)guanine activity against viruses of the herpes group. Nature 272: 583–585
Sköldenberg B, Forsgren M, Alestig K, Bergström T, Burman L, Dahlqvist E, Forkman A, Fryden A, Lövgren K, Norlin K, Norrby R, Olding-Stenkvist E, Stiernstedt G, Uhnoo I, De Vahl K (1984) Acyclovir versus vidarabine in herpes simplex encephalitis. Randomised multicentre study in consecutive Swedish patients. Lancet II: 707–711
Smith ThJ, Kremer MJ, Luo M, Vriend G, Arnold E, Kamer G, Rossmann MG, McKinlay MA, Diana GD, Otto MJ (1986) The site of attachment in human rhinovirus 14 for antiviral agents that inhibit uncoating. Science 233: 1286–1293
Tamm I, Eggers HJ (1963) Specific inhibition of replication of animal viruses. Science 142: 24–33
Whitley RJ, Soong S-J, Dolin R, Galasso GJ, Ch'ien LT, Alford CA, and the NIAID Collaborative Antiviral Study Group (1977) Adenine arabinoside therapy of biopsy-proved herpes simplex encephalitis. N Engl J Med 297: 289–294
Whitley RJ, Soong S-J, Hirsch MS, Karchmer AW, Dolin R, Galasso GJ, Dunnick JK, Alford CA (1981) Herpes simplex encephalitis: vidarabine therapy and diagnostic problems. N Engl J Med 304: 313–318
Whitley RJ, Alford CA, Hirsch MS, Schooley RT, Luby JP, Aoki FY, Hanley D, Nahmias AJ, Soong S-J, and the NIAID Collaborative Antiviral Study Group (1986) Vidarabine versus acyclovir therapy in herpes simplex encephalitis. N Engl J Med 314: 144–149

# Recombinant Vaccinia Viruses: New Applications of an Old Vaccine

Geoffrey L. Smith

Division of Virology, Department of Pathology, University of Cambridge, Tennis Court Road, Cambridge CB2 1QP, UK

## Introduction

Vaccinia virus is an orthopoxvirus that is antigenically related to variola virus, the aetiological agent of smallpox. This enabled vaccinia virus to be used as a live vaccine against smallpox, a practice that finally resulted in the eradication of this dreadful disease. Vaccinia is a large virus with a complex morphology and a double-stranded DNA genome of 185 kb (Geshelin and Berns 1974). The viral DNA is non-infectious when stripped of protein, reflecting the requirement of the virus-coded transcription al enzymes for expression of virus genes. Within the virus particle the genome is associated with many virus-coded enzymes including a complete transcription system that is able to transcribe the viral genome into functional mRNAs in infected cells without host cell protein of nucleic acid synthesis. A discrete set of early virus genes are transcribed before the onset of virus DNA replication. Some of these code for virus enzymes that are involved in further uncoating of the infecting virus core or in nucleic acid metabolism. Examples of the latter include thymidine kinase (TK) (Weir et al. 1982) and DNA polymerase (Jones and Moss 1984).

Within 2 h of infection of susceptible cells at high multiplicity, virus DNA synthesis has commenced. This occurs in the cytoplasm, as does virus transcription and assembly of virus particles. With the onset of virus DNA replication there is a marked switch in the pattern of virus transcription. Most early genes are turned off and transcription of the late virus genes is activated. Some genes are expressed throughout the virus replicative cycle. An example is the gene coding for a polypeptide of predicted molecular weight of 7.5 kd that maps within the inverted terminal repetition of the virus genome (Venkatesan et al. 1981). Although this gene is expressed throughout the virus replication cycle, it is transcribed at early and late times from different initiation sites that are separated by approximately 50 bp (Cochran et al. 1985). The late virus genes code for additional virus enzymes and the major structural components of the virus particle (Weir and Moss 1984; Wittek et al. 1984 a, b; Rosel and Moss 1985; Hirt et al. 1986; Shida 1986). During assembly of new virions discrete morphological structures can be observed with the electron microscope (Dales and Pogo 1981). In most cell culture systems routinely used for the growth of vaccinia virus only a small fraction (5%–30%) of infectious progeny virus is released from the infected cell. Therefore, virus is purified from cytoplasmic extracts of infected cells. This virus is structurally different from the extracellular released virus, the latter possessing an additional lipid envelope derived from the Golgi apparatus during morphogenesis (Payne 1980).

The mechanism(s) controlling the temporal expression of discrete sets of virus genes are being actively investigated. Nucleotide sequence analyses of virus genes and the upstream transcriptional control regions (promoters) have demonstrated that the vaccinia promoters are structurally different from those of other eukaryotic viruses or cells or of prokaryotes (Venkatesan et al. 1981, 1982; Weir and Moss 1983, 1984; Plucienniczak et al. 1985; Weinrich and Hruby 1986). Functional comparisons have also demonstrated that these promoters are transcribed only by the virus RNA polymerase and not by the RNA polymerase II of eukaryotic cells. Similarly, the vaccinia RNA polymerase does not transcribe those genes usually transcribed by the cellular RNA polymerase II (Puckett and Moss 1983).

For a more detailed review of the molecular biology of vaccinia virus the reader should refer to Moss (1985). For the purpose of describing the use of vaccinia virus as an expression vector the important biological features of the virus are: the large DNA genome that is noninfectious without associated enzymes; the presence of a virus-coded RNA polymerase; and the functionally distinct vaccinia promoters.

## Construction of Recombinant Vaccinia Viruses

Taking into account the above properties of the virus, the following strategy has been used to construct recombinant viruses (Mackett et al. 1982; Smith and Moss 1984). Firstly, to obtain expression of a foreign gene the protein coding sequences of that gene are joined to a vaccinia virus promoter. Secondly, this chimaera is inserted into the virus genome by homologous recombination in cells infected with vaccinia virus and transfected with specially constructed plasmids. The homologous recombination is facilitated by flanking the foreign gene (within a plasmid vector) with vaccinia DNA taken from a non-essential region of the virus genome. After recombination the foreign gene linked to the vaccinia promoter is inserted into the corresponding non-essential locus of the virus genome. The nature of the non-essential vaccinia DNA chosen to flank the foreign gene in the plasmid determines the site of insertion within the virus genome. Several such sites have been identified (Mackett et al. 1982; Panicali and Paoletti 1982), but a commonly used locus is the TK gene, since insertion of foreign DNA here destroys TK activity and enables the recombinant viruses to be selected as $TK^-$ mutants.

Several plasmid vectors have been constructed that are specifically designed for the transfer of genes into vaccinia virus (Mackett et al. 1984; Boyle et al. 1985; Chakrabarti et al. 1985). Alternative approaches to the selection of recombinant viruses involve expression of other selectable markers such as the neomycin resistance gene (Franke et al. 1985), the herpes simplex virus TK gene (into $TK^-$ mutants of vaccinia) (Mackett et al. 1982; Panicali and Paoletti 1982) or $\beta$-galactosidase, enabling visual selection of recombinant plaques (Chakrabarti et al. 1985).

## Applications of Recombinant Vaccinia Viruses

### The Cellular Immune Response to Virus Antigens

Several genes which code for virus glycoproteins have been inserted and expressed in recombinant vaccinia viruses. One of the first genes chosen was the influenza virus haemagglutinin (HA) (Panicali et al. 1983; Smith et al. 1983b), since this is the most important antigen for antibody-mediated neutralisation of influenza virus. It was shown that the influenza HA was synthesised in cells infected by recombinant vaccinia virus containing the HA gene and that the HA molecule was glycosylated and transported to the infected cell surface (Smith et al. 1983b). Animals immunised with the recombinant vaccinia virus produced antibodies against HA and were resistant to subsequent influenza virus infection. To determine whether the HA that was expressed on the infected cell surface could be recognised by cytotoxic T lymphocytes (CTL) in conjunction with the antigens of the major histocompatibility complex, CTL taken from mice immunised with influenza virus were used as effector cells on autologous targets infected with the recombinant vaccinia virus. The results of these $^{51}Cr$ release cytotoxicity assays clearly showed that CTL against influenza could recognise and lyse recombinant vaccinia virus-infected cells expressing HA on their surface (Bennink et al. 1984, 1986). In addition this lysis was restricted by the major histocompatibility antigen type and limited to the homologous strain of influenza HA.

Next it was tested whether the recombinant vaccinia virus expressing HA could induce an anti-HA CTL response in inoculated mice. Splenocytes taken from these mice were used as effector cells on targets infected with various strains of influenza virus. Lysis occurred only on influenza virus-infected cells in which the HA molecule was of the same subtype as the HA expressed by the recombinant vaccinia virus (Bennink et al. 1984, 1986). Therefore the anti-HA CTL were for the most part influenza A virus subtype specific.

It has been recognised for some time that the majority of CTL raised against influenza virus in mouse and man are cross-reactive among influenza A virus subtypes. Since the HA was not a major cross-reactive target antigen attention switched to the more conserved "internal" virus antigens despite the fact that it was not generally accepted that these antigens were expressed on the cell surface. Accordingly, a series of vaccinia recombinants expressing other influenza virus antigens were constructed (G. L. Smith, unpublished data) and used in cytotoxicity experiments as described above for the HA-VAC recombinant. Using the recombinant vaccinia virus expressing nucleoprotein (NP-VAC) it was shown that CTL derived from mice immunised with any of the influenza A virus subtypes could lyse target cells infected with NP-VAC, where the NP gene was derived from any H1N1 virus A/PR/8/34. The NP-VAC recombinant was also able to stimulate the production of CTL which recognised and lysed target cells infected with any influenza A virus (Yewdell et al. 1985). These data demonstrated that all or part of the NP was transported to the cell surface where it was recognised by MHC restricted CTL, and that it is a major cross-reactive target antigen for CTL derived following an influenza virus infection. Similar results have been obtained following transfection of cloned influenza virus genes into mouse cell lines (Townsend et al. 1984). These experiments illustrate the use of recombinant vaccinia viruses to study the specificity of virus antigen recognition by CTL.

## Identification of Glycoproteins with Potential as Vaccine Antigens

Vaccinia virus is a particularly useful eukaryotic expression vector system for expression of foreign glycoproteins because the polypeptide is glycosylated and transported normally within the infected cell (Smith et al. 1983a, b; Mackett et al. 1985; Stephens et al. 1986). Moreover, since the recombinant virus retains its infectivity it can be used to vaccinate experimental animals and produce monospecific antibody against the foreign gene product. This antibody may then be used to characterise the gene product and be tested for its ability to neutralise the infectivity of the pathogen from which the gene was derived. This is useful in identifying antigens from pathogens which have potential in future vaccines. Using this approach new glycoproteins of human cytomegalovirus and herpes simplex virus type I have been identified which induce neutralising antibodies against the respective viruses (unpublished data).

## Live Vaccines

Since vaccinia virus has been established as an effective live vaccine the possibility exists that recombinant vaccinia viruses expressing foreign genes from other pathogens may be used as live vaccines against the appropriate pathogen. It has been demonstrated that vaccination with recombinant vaccinia viruses induces immunity that protects against several eukaryotic viruses. These include influenza virus (Smith et al. 1983b; Small et al. 1985), herpes simplex virus (Paoletti et al. 1984; Cremer et al. 1985), hepatitis B virus (Moss et al. 1984), rabies (Kieny et al. 1984; Wiktor et al. 1984), vesicular stomatitis virus (VSV) (Mackett et al. 1985) and respiratory syncytial virus (Elango et al. 1986). This type of live recombinant vaccine has many attractions. Several of these were also important features in the effectiveness of vaccinia virus as a vaccine against smallpox. (1) The virus produces a live infection in the vaccinated host and is able to induce both humoral antibody and cellular immune responses. (2) The vaccinia virus has a large capacity for foreign DNA (Smith and Moss 1983) and this permits the simultaneous expression of multiple foreign genes so creating a polyvalent vaccine against multiple pathogens (Perkus et al. 1985). (3) The vaccine is relatively simple and economical to manufacture in either animal skin or tissue culture. This is increasingly important in poorer countries in which the extremely limited health budgets do not permit the purchase of expensive vaccines. (4) The vaccine is simple to administer in non-sterile field conditions by relatively unskilled personnel. Previously this was done using a bifurcated needle or jet gun. (5) The freeze-dried vaccine is extremely stable without refrigeration. This has enabled transport in tropical countries without the difficult and expensive task of maintaining a "cold chain". (6) Vaccinia virus has a broad host range permitting its potential use as vaccines against medical and veterinary pathogens. The protection of cattle against VSV is a good example of the potential applications in veterinary medicine.

Although these advantages are impressive there are also problems with the re-introduction of vaccinia virus as an immunising agent. During the smallpox eradication campaign it was recognised that there were rare but serious complications following vaccination. While these rare complications were relatively unimportant from a public health point of view in the face of a life-threatening disease, with the disappearance of smallpox they were naturally a major concern and a good reason for discontinuing vaccination. Therefore, if immunisation with vaccinia virus is to be reintroduced, it is necessary to construct more attenuated vaccinia viruses via genetic manipulation to reduce and hopefully eliminate these troublesome complications. Towards this goal, genes involved in the pathogenicity of vaccinia virus are beginning to be identified. The inactivation of one of these (TK) has been shown to markedly attenuate the virus (Buller et al. 1985).

A second potential problem with vaccination using recombinant vaccinia viruses is the potential to change the pathogenicity or tissue tropism of the recombinant virus due to the expression of the foreign gene. For instance, the expression in vaccinia virus of a surface glycoprotein from a neurotropic pathogen such as rabies or a lymphotropic pathogen such as AIDS virus could conceivably change vaccinia from a dermotropic to a neurotropic or lymphotropic virus. While there are no data indicating that this does in fact occur, it is a potential problem that must be evaluated for each foreign gene.

Lastly, a large proportion of the world's population possesses some degree of immunity to vaccinia owing to previous smallpox vaccinations. This immunity will naturally diminish or prevent virus replication upon subsequent revaccination and thereby reduce the level of the immune response to new antigens expressed in the recombinant virus. However, this problem of existing immunity to vaccinia may not be as great as previously thought, for the following reasons. First, since smallpox vaccination has been discontinued in many countries for up to 10 years there is an ever-growing population of children who have not been vaccinated with vaccinia virus. Secondly, it was recognised during smallpox vaccination campaigns that repeat vaccinations do "take" and cause a boost to existing immunity. Using experimental animals it has also been shown that repeat vaccinations with recombinant viruses can cause boosting in antibody levels against the foreign gene product. An accidental vaccination of a human with a recombinant vaccinia virus expressing VSV nucleocapsid protein resulted in antibody production against the VSV protein despite the person having previously had a smallpox vaccination (Jones et al. 1986). Clearly, revaccination can be effective. It is also worth remembering that many diseases are first experienced in childhood and so vaccination is required early in life. As pointed out, children possess no immunity to vaccinia virus.

## Summary

Vaccinia virus is the world's oldest vaccine that was used in the global eradication of smallpox. Genetic engineering has enabled recombinant vaccinia viruses that express foreign genes to be constructed. These recombinant vaccinia viruses have several applications including new live vaccines, tools for studying the immune response to individual antigens and a way of studying the regulation of vaccinia gene expression.

*Acknowledgement.* I would like to express my thanks to Mary Wright to typing the manuscript.

## References

Bennink JR, Yewdell JW, Smith GL, Moller C, Moss B (1984) Recombinant vaccinia virus primes and stimulates influenza virus haemagglutinin-specific cytotoxic T lymphocytes. Nature 311: 578–579

Bennink JR, Yewdell JW, Smith GL, Moss B (1986) Recognition of cloned influenza virus haemagglutinin gene products by cytotoxic T lymphocytes. J Virol 57: 786–791

Boyle DB, Couper BEH, Both GW (1985) Multiple-cloning-site plasmids for the rapid construction of recombinant poxviruses. Gene 35: 169–177

Buller RML, Smith GL, Cremer K, Notkins AL, Moss B (1985) Decreased virulence of recombinant vaccinia virus expression vectors is associated with a thymidine kinase-negative phenotype. Nature 317: 813–815

Chakrabarti S, Brechling K, Moss B (1985) Vaccinia virus expression vector: co-expression of $\beta$-galactosidase provides visual screening of recombinant virus plaques. Mol Cell Biol 5: 3403–3409

Cochran MA, Puckett C, Moss B (1985) In vitro mutagenesis of the promoter region of a vaccinia virus gene: evidence for tandem early and late regulatory signals. J Virol 54: 30–37

Cremer K, Mackett M, Wohlenberg C, Notkins AL, Moss B (1985) Vaccinia virus recombinants expressing herpes simplex virus type 1 glycoprotein D prevent latent herpes in mice. Science 228: 737–740

Dales S, Pogo BT (1981) Biology of poxviruses. Springer, Berlin Heidelberg New York, pp 1–109 (Virology monographs, vol 18)

Elango N, Prince GA, Murphy BR, Venkateran S, Chanock RM, Moss B (1986) Resistance to human respiratory syncytial virus (RSV) infection induced by immunisation of cotton rats with a recombinant vaccinia virus expressing the RSV G glycoprotein. Proc Natl Acad Sci USA 83: 1906–1910

Franke CA, Rice CM, Strauss JH, Hruby DE (1985) Neomycin resistance as a dominant selectable marker for selection and isolation of vaccinia virus recombinants. Mol Cell Biol 5: 1918–1924

Geshelin P, Berns KI (1974) Characterization and localisation of the naturally occurring cross-links in vaccinia virus DNA. J Mol Biol 88: 785–796

Hirt P, Hiller G, Wittek R (1986) Localization and fine structure of a vaccinia virus gene encoding an envelope antigen. J Virol 58: 757–764

Jones EV, Moss B (1984) Mapping of the vaccinia virus DNA polymerase gene by marker rescue and by cell-free translation of selected RNA. J Virol 49: 72–77

Jones L, Ristow S, Yilma T, Moss B (1986) Accidental human vaccination with vaccinia virus expressing nucleoprotein gene. Nature 319: 543

Kieny MP, Lathe R, Drillien R, Spehner D, Skory S, Schmitt D, Wiktor T, Koprowski H, Lecocq JP (1984) Expression of rabies virus glycoprotein from a recombinant vaccinia virus. Nature 312: 163–166

Mackett M, Smith GL, Moss B (1982) Vaccinia virus: a selectable eukaryotic cloning and expression vector. Proc Natl Acad Sci USA 79: 7415–7419

Mackett M, Smith GL, Moss B (1984) General method for production and selection of infectious vaccinia virus recombinants expressing foreign genes. J Virol 49: 857–864

Mackett M, Yilma T, Rose J, Moss B (1985) Vaccinia virus recombinants: expression of VSV genes and protective immunization of mice and cattle. Science 227: 433–435

Moss B (1985) Replication of poxviruses. In: Fields BN, Chanock RM, Roizman B (eds) Virology. Raven, New York, pp 658–703

Moss B, Smith GL, Gerin JL, Purcell RH (1984) Live recombinant vaccinia virus protects chimpanzees against hepatitis B. Nature 311: 67–69

Panicali D, Paoletti E (1982) Construction of poxviruses as cloning vectors: insertion of the thymidine kinase from herpes simplex virus into the DNA of infectious vaccinia virus. Proc Natl Acad Sci USA 79: 4927–4931

Panicali D, Davis SW, Weinberg RL, Paoletti E (1983) Construction of live vaccines by using genetically engineered poxviruses: biological activity of recombinant vaccinia virus expressing influenza virus haemagglutinin. Proc Natl Acad Sci USA 80: 5364–5368

Paoletti E, Lipinskas BR, Samsanoff C, Mercer S, Panicali D (1984) Construction of live vaccines using genetically engineered poxviruses: biological activity of vaccinia virus recombinants expressing the hepatitis B virus surface antigen and the herpes simplex virus glycoprotein D. Proc Natl Acad Sci USA 81: 193–197

Payne LG (1980) Significance of extracellular enveloped virus in the in vitro and in vivo dissemination of vaccinia. J Gen Virol 50: 89–100

Perkus ME, Piccini A, Lipinskas BR, Paoletti E (1985) Recombinant vaccinia virus: immunization against multiple pathogens. Science 229: 981–984

Plucienniczak A, Schroeder E, Zettlmeissl G, Streeck RE (1985) Nucleotide sequence of a cluster of early and late genes in a conserved segment of the vaccinia virus genome. Nucleic Acids Res 13: 985–998

Puckett C, Moss B (1983) Selective transcription of vaccinia virus genes in template-dependent extracts of infected cells. Cell 35: 441–448

Rosel J, Moss B (1985) Transcriptional and translational mapping and nucleotide sequence of a vaccinia virus gene encoding the precursor of the major core polypeptide 4b. J Virol 56: 830–838

Shida H (1986) Nucleotide sequence of the vaccinia virus hemagglutinin gene. Virology 150: 451–462

Small PA Jr, Smith GL, Moss B (1985) Intranasal vaccination with recombinant vaccinia containing influenza haemagglutinin prevents both influenza virus pneumonia and nasal infection: intradermal vaccination prevents only viral pneumonia. In: Quinnan GV (ed) Vaccinia viruses as vectors for vaccine antigen. Elsevier, New York, pp 175–178

Smith GL, Moss B (1983) Infectious poxvirus vectors have capacity for at least 25000 base pairs of foreign DNA. Gene 25: 21–28

Smith GL, Moss B (1984) Vaccinia virus expression vectors: construction, properties and application. Biotechniques 2: 306–312

Smith GL, Mackett M, Moss B (1983a) Infectious vaccinia virus recombinants that express hepatitis B virus surface antigen. Nature 302: 490–495

Smith GL, Murphy BR, Moss B (1983b) Construction and characterization of an infectious vaccinia virus recombinant that expresses the influenza virus haemagglutinin gene and induces resistance to influenza infection in hamsters. Proc Natl Acad Sci USA 80: 7155–7159

Stephens EB, Compans RW, Earl P, Moss B (1986) Surface expression of viral glycoproteins is polarized in epithelial cells infected with recombinant vaccinia viral vectors. EMBO J 5: 237–245

Townsend ARM, McMichael AJ, Carter NP, Huddleston JA, Brownlee GG (1984) Cytotoxic T cell recognition of the influenza nucleoprotein and haemagglutinin expressed in transfected L cells. Cell 39: 13–25

Venkatesan S, Baroudy BM, Moss B (1981) Distinctive nucleotide sequences adjacent to multiple initiation and termination sites of an early vaccinia virus gene. Cell 25: 805–813

Venkatesan S, Gershowitz A, Moss B (1982) Complete nucleotide sequence of two adjacent early vaccinia virus genes located within the inverted terminal repetition. J Virol 44: 637–646

Weinrich SL, Hruby DE (1986) A tandemly-oriented late gene cluster within the vaccinia virus genome. Nucleic Acids Res 14: 3003–3016

Weir JP, Moss B (1983) Nucleotide sequence of the vaccinia virus thymidine kinase gene and the nature of spontaneous frameshift mutations. J Virol 46: 530–537

Weir JP, Moss B (1984) Regulation of expression and nucleotide sequence of a late vaccinia virus gene. J Virol 51: 662–669

Weir JP, Bajszar G, Moss B (1982) Mapping of the vaccinia virus thymidine kinase gene by marker rescue and cell-free translation of selected mRNA. Proc Natl Acad Sci USA 79: 1210–1214

Wiktor TJ, Macfarland RI, Reagan KJ, Dietzschold B, Curtis PJ, Wunner WH, Kieny M-P, Lathe R, Lecocq J-P, Mackett M, Moss B, Koprowski H (1984) Protection from rabies by a vaccinia virus recombinant containing the rabies virus glycoprotein gene. Proc Natl Acad Sci USA 81: 7194–7198

Wittek R, Hanaggi M, Hiller G (1984a) Mapping of a gene coding for a major late structural polypeptide on the vaccinia virus genome. J Virol 49: 371–378
Wittek R, Richner B, Hiller G (1984b) Mapping of the genes coding for the major vaccinia core polypeptides. Nucleic Acids Res 12: 4835–4848
Yewdell JW, Bennink JR, Smith GL, Moss B (1985) Influenza A virus nucleoprotein is a major target antigen for cross-reactive anti-influenza A virus cytotoxic T lymphocytes. Proc Natl Acad Sci USA 82: 1785–1789

# Subject Index

MIX
Papier aus verantwortungsvollen Quellen
Paper from responsible sources
FSC® C105338

If you have any concerns about our products,
you can contact us on
**ProductSafety@springernature.com**

In case Publisher is established outside the EU,
the EU authorized representative is:
**Springer Nature Customer Service Center GmbH**
**Europaplatz 3, 69115 Heidelberg, Germany**

Printed by Libri Plureos GmbH
in Hamburg, Germany